THE DECADE OF DISCOVERY IN ASTRONOMY AND ASTROPHYSICS

Astronomy and Astrophysics Survey Committee
Board on Physics and Astronomy
Commission on Physical Sciences, Mathematics, and Applications
National Research Council

NATIONAL ACADEMY PRESS
Washington, D.C. 1991

NATIONAL ACADEMY PRESS • 2101 Constitution Avenue, NW • Washington, DC 20418

NOTICE: The project that is the subject of this report was approved by the Governing Board of the National Research Council, whose members are drawn from the councils of the National Academy of Sciences, the National Academy of Engineering, and the Institute of Medicine. The members of the committee responsible for the report were chosen for their special competences and with regard for appropriate balance.

This report has been reviewed by a group other than the authors according to procedures approved by a Report Review Committee consisting of members of the National Academy of Sciences, the National Academy of Engineering, and the Institute of Medicine.

This project was supported by the Department of Energy under Grant No. DE-FG05-89ER40421, the National Aeronautics and Space Administration and the National Science Foundation under Grant No. AST-8901685, the Naval Research Laboratory under Contract No. N00173-90-M-9744, and the Smithsonian Institution under Purchase Order No. SF0022430000. Additional support was provided by the Maurice Ewing Earth and Planetary Sciences Fund of the National Academy of Sciences created through a gift from the Palisades Geophysical Institute, Inc., and an anonymous donor.

Library of Congress Cataloging-in-Publication Data
National Research Council (U.S.). Astronomy and Astrophysics Survey Committee.
The decade of discovery in astronomy and astrophysics / Astronomy and Astrophysics Survey Committee, Board on Physics and Astronomy, Commission on Physical Sciences, Mathematics, and Applications, National Research Council.
p. cm.
Includes bibliographical references and index.
ISBN 0-309-04381-6
1. Astronomy—Research. 2. Astrophysics—Research. I. Title.
QB61.N38 1991
520′.72—dc20 90-21659
CIP

Cover: A view of the Milky Way Galaxy obtained by NASA's Cosmic Background Explorer (COBE). The central parts of the Milky Way, the galaxy in which our sun and solar system are located, are normally obscured by intervening gas and dust. Observations in the near-infrared reveal a thin disk and a central bulge of stars at the center of the galaxy, located some 28,000 light-years away. Courtesy of the COBE Science Working Group and NASA's Goddard Space Flight Center.

Printed in the United States of America

NATIONAL RESEARCH COUNCIL

2101 CONSTITUTION AVENUE WASHINGTON, D. C. 20418

OFFICE OF THE CHAIRMAN

This report on astronomy and astrophysics, *The Decade of Discovery*, is both timely and unique. During a period when the federal government recognizes the need for, and the difficulty of, fiscal restraint, the report sets clear priorities for funding. The report is unusual in the widespread sponsorship by governmental agencies seeking advice on the research they sponsor in this field and in the breadth of participation by the relevant professional community.

The report describes a prioritized set of research initiatives that excite the intellect and stir a sense of adventure, ranging from understanding the large-scale distribution of matter in the universe to searching for planetary systems around nearby stars. In recent years, remarkable discoveries about the universe have attracted professional scientists trained in other fields to astrophysics and have stimulated many young people to study scientific or technical subjects. This study documents the amazing improvements in our ability to investigate the heavens made possible by recent advances in telescopes, detectors, and computers. At the same time, it makes clear the urgent need to maintain and improve the nation's infrastructure for research in astronomy.

In the organizational stages of the committee's work, the survey's chair visited, together with the chair of the National Research Council or with an NRC staff representative, leaders of the sponsoring agencies, members of Congress and their staffs, and relevant individuals in the Office of Management and Budget. These visits helped astronomers direct their efforts to questions pertinent to governmental needs, while preserving the ability of the committee to make independent judgments based on scientific expertise. Parts of this report address questions raised in those early consultations, including chapters on the potential of astronomy from the moon; on education, science management, and international collaborations; and on the importance of astronomy to broad societal goals.

The committee considered many more equipment initiatives than it could recommend with fiscal responsibility. Thus the committee prioritized initiatives on the basis of their scientific importance, timeliness, and cost-effectiveness. The widely inclusive nature of the committee's deliberations ensures that this prioritization will have the support of the great majority of the nation's astronomers. The underlying strength of this report is the excitement of the recent discoveries that are surveyed and the clear path to even more extraordinary revelations that is set forth in the new initiatives.

The arguments presented in favor of these initiatives include the universal appeal of astronomy to the curious of all ages, the stimulating educational effects of astronomical programs, the strong linkage with other physical sciences, and the remarkable but unforeseen applications of astronomical techniques to more practical endeavors.

I would like to thank the Astronomy and Astrophysics Survey Committee and the panel members for the large investment of time and energy represented in this report. The fortunes of astronomy in any nation are symbolic of the treatment of all of basic science. John Bahcall understood this, and he has earned the gratitude of the scientific community as a whole and astronomers in particular for his inspired leadership and vision in developing an extraordinary consultative process and bringing this survey to a responsible conclusion.

Frank Press
Chair
National Research Council

ASTRONOMY AND ASTROPHYSICS SURVEY COMMITTEE

JOHN N. BAHCALL, Institute for Advanced Study, *Chair*
CHARLES A. BEICHMAN, Institute for Advanced Study, *Executive Secretary*
CLAUDE CANIZARES, Massachusetts Institute of Technology
JAMES CRONIN, University of Chicago
DAVID HEESCHEN, National Radio Astronomy Observatory
JAMES HOUCK, Cornell University
DONALD HUNTEN, University of Arizona
CHRISTOPHER F. McKEE, University of California, Berkeley
ROBERT NOYES, Harvard-Smithsonian Center for Astrophysics
JEREMIAH P. OSTRIKER, Princeton University Observatory
WILLIAM PRESS, Harvard-Smithsonian Center for Astrophysics
WALLACE L.W. SARGENT, California Institute of Technology
BLAIR SAVAGE, University of Wisconsin
ROBERT W. WILSON, AT&T Bell Laboratories
SIDNEY WOLFF, National Optical Astronomy Observatories

National Research Council Staff

Robert L. Riemer, Senior Program Officer
Susan M. Wyatt, Administrative Associate
  Board on Physics and Astronomy
William Spindel, Principal Staff Officer (1989)
Sandra Nolte, Administrative Assistant (1989-1990)
Phoebe Wechsler, Administrative Assistant (1989-1990)
  Institute for Advanced Study

BOARD ON PHYSICS AND ASTRONOMY

COMMISSION ON PHYSICAL SCIENCES, MATHEMATICS, AND APPLICATIONS*

NORMAN HACKERMAN, Robert A. Welch Foundation, *Chairman*
PETER J. BICKEL, University of California, Berkeley
GEORGE F. CARRIER, Harvard University
HERBERT D. DOAN, The Dow Chemical Company (retired)
DEAN E. EASTMAN, IBM T.J. Watson Research Center
MARYE ANNE FOX, University of Texas
PHILLIP A. GRIFFITHS, Duke University
NEAL F. LANE, Rice University
ROBERT W. LUCKY, AT&T Bell Laboratories
CHRISTOPHER F. McKEE, University of California, Berkeley
RICHARD S. NICHOLSON, American Association for the Advancement of Science
JEREMIAH P. OSTRIKER, Princeton University Observatory
ALAN SCHRIESHEIM, Argonne National Laboratory
ROY F. SCHWITTERS, Superconducting Super Collider Laboratory
KENNETH G. WILSON, Ohio State University

NORMAN METZGER, Executive Director

*The project that is the subject of this report was initiated by the predecessor group of the Commission on Physical Sciences, Mathematics, and Applications, which was the Commission on Physical Sciences, Mathematics, and Resources, whose members are listed in Appendix D.

The National Academy of Sciences is a private, nonprofit, self-perpetuating society of distinguished scholars engaged in scientific and engineering research, dedicated to the furtherance of science and technology and to their use for the general welfare. Upon the authority of the charter granted to it by the Congress in 1863, the Academy has a mandate that requires it to advise the federal government on scientific and technical matters. Dr. Frank Press is president of the National Academy of Sciences.

The National Academy of Engineering was established in 1964, under the charter of the National Academy of Sciences, as a parallel organization of outstanding engineers. It is autonomous in its administration and in the selection of its members, sharing with the National Academy of Sciences the responsibility for advising the federal government. The National Academy of Engineering also sponsors engineering programs aimed at meeting national needs, encourages education and research, and recognizes the superior achievements of engineers. Dr. Robert M. White is president of the National Academy of Engineering.

The Institute of Medicine was established in 1970 by the National Academy of Sciences to secure the services of eminent members of appropriate professions in the examination of policy matters pertaining to the health of the public. The Institute acts under the responsibility given to the National Academy of Sciences by its congressional charter to be an adviser to the federal government and, upon its own initiative, to identify issues of medical care, research, and education. Dr. Samuel O. Thier is president of the Institute of Medicine.

The National Research Council was established by the National Academy of Sciences in 1916 to associate the broad community of science and technology with the Academy's purposes of furthering knowledge and of advising the federal government. Functioning in accordance with general policies determined by the Academy, the Council has become the principal operating agency of both the National Academy of Sciences and the National Academy of Engineering in providing services to the government, the public, and the scientific and engineering communities. The Council is administered jointly by both Academies and the Institute of Medicine. Dr. Frank Press and Dr. Robert M. White are chairman and vice chairman, respectively, of the National Research Council.

Preface

The National Research Council commissioned the Astronomy and Astrophysics Survey Committee, a group of 15 astronomers and astrophysicists, to survey their field and to recommend new ground- and space-based programs for the coming decade. Support was provided by the National Science Foundation, the National Aeronautics and Space Administration, the Department of Energy, the United States Navy, and the Smithsonian Institution.

The survey committee's chair was appointed in February 1989 by the president of the National Academy of Sciences upon recommendation of a committee selected by the Commission on Physical Sciences, Mathematics, and Resources. The chair sent a letter to all members of the astronomy section of the National Academy, to the chairs of all astronomy departments in the United States, and to other leading astronomers inviting nominations for individuals to serve on the Astronomy and Astrophysics Survey Committee. The committee members were selected by the Board on Physics and Astronomy and appointed by the National Research Council after extensive discussions by the chair with interested astronomers.

The survey committee established 15 advisory panels to represent different wavelength subdisciplines, as well as solar, planetary, theoretical, and laboratory astrophysics. The chairs of the subdiscipline panels helped the committee to select a broad and representative group of experts, totaling more than 300 people. The panel chairs were responsible, together with their panel members, for obtaining the views of a wide cross-section of the astronomy and astrophysics community and for preparing a paper on their discussions and findings. A member of the survey committee served as a vice-chair of each panel.

Ten panels had charges that reflected specific scientific areas, eight of them based on wavelength region and two (those of the Planetary Astronomy Panel and Solar Astronomy Panel) on particular subdisciplines with special needs. The committee asked these ten science panels to identify the most important scientific goals in their respective areas, to prioritize the new initiatives needed to achieve these goals, to recommend proposals for technology development, to consider the possibilities for international collaboration, and to discuss any policy issues relevant to their charge. The Astronomy and Astrophysics Survey Committee served as an interdisciplinary panel to guarantee that scientific questions that did not fit conveniently into this organizational structure were handled appropriately on an ad hoc basis.

Four other panels were appointed to explore computing and data processing, policy opportunities, the benefits of astronomy to the nation, and the status of the profession. The working papers written on the first three topics were used by the committee as a basis for developing the chapters with corresponding subject matter (Chapters 5, 7, and 8, respectively) in the survey report. Data from the working paper titled "Status of the Profession" were used in preparing various chapters and Appendix B of the survey report and by other panels in preparing their papers. The Science Opportunities Panel, the fifteenth panel appointed by the committee, prepared a paper that the committee believed should be expanded and published separately as a popular book accessible to as large an audience as possible. An abbreviated and adapted version of this panel's paper appears as Chapter 2 of the survey report. The Lunar Working Group of the committee prepared a paper that appears as Chapter 6, "Astronomy from the Moon," in this report.

Members of the panels consulted widely with their colleagues to solicit advice and to inform other members of the astronomical community of the main issues facing the committee. Each panel held an open meeting at a session of the American Astronomical Society, and most of the panels held sessions at other professional gatherings, as well as at astronomical centers at different places in the United States. Each panel discussed with the relevant federal agency personnel the problems and issues of its particular area. These interactions with agency personnel provided valuable background to the discussions, although the panels were careful to preserve the independence and confidentiality of the National Research Council deliberative process.

The panel chairs presented their papers in oral and written form at the June and July 1990 meetings of the survey committee and were invited to participate with the committee in the initial attempts to generate a cohesive set of overall recommendations. The views of the participants were modified by the discussions that took place between the different advocates and experts. The committee based its final decisions and recommendations in significant part on the contents of the panel papers and on the discussions with the panel chairs.

The unrefereed working papers of the subdiscipline panels give technical details about many of the programs discussed in this report. They are contained in the separately published *Working Papers: Astronomy and Astrophysics Panel Reports* (National Academy Press, Washington, D.C., 1991) issued by the National Research Council. These papers were advisory to the survey committee and represent the opinions of the members of each panel in the context of their individual charges.

The committee is grateful to the many other astronomers, both in the United States and from abroad, who provided written advice or participated in organized discussions. In all, more than 15 percent of American astronomers played an active role in at least one aspect of this study. Appendix C lists the members of the subdiscipline panels.

Many other people too numerous to cite individually assisted in various aspects of the survey. The committee gratefully acknowledges Rebecca Elson as technical editor, Susan Maurizi as general editor, and Margaret Best and Phoebe Wechsler for their invaluable efforts in preparing the seemingly infinite number of drafts of this report. R. Riemer provided guidance and support to the committee in his capacity as staff officer. C. Beichman served effectively in dual roles as Executive Secretary and as a member of the committee. Finally, the survey's chair thanks Frank Press for generous doses of his wisdom and insight during the past two years.

JOHN BAHCALL
Chair
Astronomy and Astrophysics
Survey Committee

Contents

APPENDICES

THE DECADE OF DISCOVERY IN ASTRONOMY AND ASTROPHYSICS

Executive Summary

THE DECADE OF DISCOVERY

In the coming decade astronomers will use telescopes in space, in aircraft, on the ground, and even underground to address fundamental questions concerning our place in the universe. Do planets orbit nearby stars? What triggers the formation of stars? How do life-giving elements such as carbon and oxygen form and disperse throughout the galaxy? Where can black holes be found, and do they power luminous galaxies and quasars? How and when did galaxies form? Will the universe continue to expand forever, or will it reverse its course and collapse on itself?

The 1990s promise to be a decade of discovery, providing at least partial answers to known fundamental questions. New instruments will reveal previously unimagined aspects of the universe and will lead to new questions about objects that we do not yet know exist.

The National Research Council commissioned the Astronomy and Astrophysics Survey Committee, a group of 15 astronomers and astrophysicists, to survey their field and to recommend the most important new ground- and space-based initiatives for the coming decade. The survey committee obtained advisory studies from over 300 astronomers who participated in one or more of the 15 panels established to represent different wavelength disciplines, as well as solar, planetary, theoretical, and laboratory astrophysics. Many other astronomers provided written advice or participated in organized discussions. More than 15 percent of American astronomers played an active role in some aspect of this study. The committee also consulted with distinguished foreign scientists on future directions in astronomy.

In addition to constructing a prioritized list of new instruments based on its assessment of the opportunities for fundamental scientific advances, the committee also evaluated the existing infrastructure, considered the human aspects of the field, including education and international collaborations, explored the consequences of the computer revolution for astronomy, investigated the astronomical opportunities provided by lunar observatories, prepared a popular summary of opportunities for scientific advances in astronomy, and suggested the most promising areas for developing new observational technologies. The committee also examined the ways in which astronomical research contributes to society.

For the decade of the 1990s, the committee places the highest priority for ground-based research on increased support for the infrastructure for astronomy and sets as the highest priority for space-based research the establishment of a program that has an appropriate balance between more frequent small and moderate projects and unique large projects. The prioritized list of new equipment initiatives primarily reflects the committee's assessment of the relative scientific potential of the different projects. The committee also took into account cost-effectiveness, technological readiness, educational impact, and the relation of each project to existing or proposed initiatives in the United States and in other countries.

RESTORING THE INFRASTRUCTURE

The committee's highest priority for ground-based research is to strengthen the research infrastructure at universities and at the national observatories.

- **The committee recommends that the National Science Foundation increase the operations and maintenance budgets of the national observatories to an adequate and stable fraction of their capital cost, thereby repairing the damage caused by a decade of deferred maintenance.**

- **The committee recommends an increase in the individual grants program in astronomy at the NSF to permit young researchers to take advantage of the new opportunities for discovery, to utilize appropriately the large amounts of new data, and to enhance support for theoretical astrophysics.**

ACHIEVING A BALANCED SPACE PROGRAM

For space astronomy, the report highlights the need for a balanced program that includes both the Great Observatories and more frequent, smaller missions.

The Great Observatories are large facilities that make possible "small science" at institutions distributed across the country, since typically only a few researchers work on each observing project. The committee reexamined the justification for large-scale space astronomy programs, taking into account both the failure to meet specifications for the Hubble Space Telescope and the National Aeronautics and Space Administration's (NASA's) record of successes in carrying out other complex missions at the frontiers of science and technology.

The committee concluded that large telescopes are required to answer some of the most fundamental questions in astronomy. However, smaller telescopes can be built and launched more quickly to answer specific questions, to respond to technological innovations, and to train future generations of scientists.

THE PRIORITIZED INSTRUMENTAL PROGRAM

Progress in astronomy often comes from technological advances that open new windows on the universe or make possible large increases in sensitivity or resolution. During the 1990s, arrays of infrared detectors, the ability to build large optical telescopes, improved angular resolution at a variety of wavelengths, new electronic detectors, and the ability of computers to process large amounts of data will make possible an improved view of the universe.

Paralleling these breakthroughs in observational capability are greater demands on astrophysical theory that are being met with new generations of computers. Modern instruments reveal details that often require sophisticated models for their interpretation. Successfully comparing the results of observations with the constructs of theory, and predicting new phenomena, will require deeper understanding of astrophysical processes, more clever algorithms, more computational power, and improved knowledge of physical constants.

The survey committee divided its recommendations for instrumental initiatives into categories of large, moderate, and small programs, depending on the scale of the necessary resources. Ground- and space-based facilities were prioritized in a combined list and also separately.

Table 1 presents the recommended list of ground- and space-based equipment initiatives in the large and moderate-sized categories. The four large programs recommended for construction in the 1990s are described below in order of priority.

Large Programs

- The **Space Infrared Telescope Facility (SIRTF)**, which would complete NASA's Great Observatory program, would be almost a thousand times more sensitive than earth-based telescopes operating in the infrared. Advanced arrays of infrared detectors, pioneered in the United States, would give SIRTF the ability to map complex areas and measure spectra a million times faster

TABLE 1 Recommended Equipment Initiatives (Combined Ground and Space) and Estimated Costs

Initiative	Decade Cost ($M)
Large Programs	
Space Infrared Telescope Facility (SIRTF)	1,300
Infrared-optimized 8-m telescope	80
Millimeter Array (MMA)	115
Southern 8-m telescope	55
Subtotal for large programs	1,550
Moderate Programs	
Adaptive optics	35
Dedicated spacecraft for FUSE	70
Stratospheric Observatory for Far-Infrared Astronomy (SOFIA)	230
Delta-class Explorer acceleration	400
Optical and infrared interferometers	45
Several shared 4-m telescopes	30
Astrometric Interferometry Mission (AIM)	250
Cosmic-ray telescope (Fly's Eye)	15
Large Earth-based Solar Telescope (LEST)	15
VLA extension	32
International collaborations on space instruments	100
Subtotal for moderate programs	1,222
Subtotal for small programs[a]	251
DECADE TOTAL	3,023

[a] See Chapter 1 for details.

than any other space-borne infrared telescope. Two successful Explorer missions provide an excellent technical heritage for SIRTF.

- An **infrared-optimized 8-m U.S. telescope** operating on Mauna Kea, Hawaii, would provide a unique and powerful instrument for studying the origin, structure, and evolution of planets, stars, and galaxies. With diffraction-limited angular resolution better than a tenth of an arcsecond, high sensitivity due to the low telescope background, and instruments capable of high spectral resolution, the infrared-optimized 8-m telescope would complement SIRTF across the limited range of wavelengths transmitted by the atmosphere. Plans for this telescope draw on a decade of progress in the technology of building large mirrors.
- The **Millimeter Array (MMA)**, an array of telescopes operating at millimeter wavelengths, would provide high-spatial- and high-spectral-resolution images of star-forming regions and distant star-burst galaxies. With spatial resolution of a tenth of an arcsecond at a wavelength of 1 mm, the MMA

would bring new classes of objects into clear view for the first time. The MMA utilizes experience and technology developed for the Very Large Array and for two smaller millimeter arrays.

- An **8-m optical telescope**, operating from the Southern Hemisphere, would give U.S. astronomers access to important objects in southern skies. All-sky coverage is essential for pursuing many of the most fundamental astronomical questions.

Small and Moderate Programs

Small and moderate-sized programs can be carried out relatively quickly in response to new scientific or technological developments, focusing research into the currently most rewarding areas and making possible greater participation by young astronomers. Some of the most exciting scientific results of the past decade have come from modest, cost-effective programs.

- **The committee recommends that an increased emphasis be given in the astronomy research budget to small and moderate programs.**

SPACE-BASED PROGRAMS

Recommendations for moderate-sized space programs include a three-phase augmentation of NASA's Explorer program by purchasing a dedicated spacecraft for the Far Ultraviolet Spectroscopy Explorer (FUSE), increasing the number of astrophysics missions launched on Delta rockets to six for the decade, and increasing the number of astrophysics Explorers launched on Scout-class rockets to five for the decade.

The Stratospheric Observatory for Far-Infrared Astronomy (SOFIA), a moderate-sized telescope in a 747 aircraft, would open submillimeter and far-infrared wavelengths to routine observation and would help train new generations of experimentalists. The committee emphasizes that a moderate-class Astrometric Interferometry Mission (AIM) capable of measuring the positions of astronomical objects with a precision of a few millionths of an arcsecond would have a great impact on many branches of astronomy. The committee recommends specific funding for flying U.S. instruments on foreign spacecraft.

GROUND-BASED PROGRAMS

For ground-based astronomy, the committee judged that two innovative techniques, adaptive optics and interferometry, can greatly enhance the spatial resolution of astronomical images. Adaptive optics can ameliorate the distorting effects of atmospheric turbulence and can be applied to existing or planned telescopes at infrared wavelengths, and eventually at optical wavelengths. The

Large Earth-based Solar Telescope (LEST) would provide important information about the sun and test the application of adaptive optics techniques. Optical and infrared interferometry promises spatial resolution better than a thousandth of an arcsecond by linking the outputs of widely separated telescopes. Improvements in the quality and spatial resolution of radio images will be possible with an extension to the Very Large Array.

The committee urges the construction of additional 4-m optical telescopes to provide greater access for U.S. scientists to state-of-the-art instrumentation capable of addressing significant astronomical problems. Private and state funds should be sufficient to build and operate such instruments, augmented with modest federal assistance. The committee urges construction of an innovative telescope to determine the characteristics of the most energetic cosmic rays. The committee emphasizes the importance of continued funding for the development of improved detectors at optical and infrared wavelengths, and of instruments to detect neutrinos and "dark matter."

THEORY AND COMPUTERS

The success of modern astrophysics illustrates the close interdependence of theory, observations, and experiment. The committee believes that NSF and NASA should increase their support for relatively inexpensive theoretical and laboratory work that is crucial to the interpretation of the results from major observatories.

- **The committee recommends that theoretical astrophysics be funded as a separate program and be given additional resources within the NSF. Within NASA, support for theoretical astrophysics should grow in approximate proportion to the support for the analysis and interpretation of observational data.**

Astronomers use computers to collect and study billions of bytes of data every 24 hours and to make theoretical simulations of complex phenomena.

- **The committee recommends establishment of national electronic archives of ground- and space-based data, the purchase of desktop and departmental computers, and the development of fast networks to link most astronomical computers together.**

LUNAR ASTRONOMY

A major space exploration initiative could return humans to the moon in the early part of the next century. The committee studied the suitability of the moon for possible astronomical facilities and concluded that a cost-effective and scientifically productive program would require early technology development, including pilot programs with substantial scientific return.

- **The committee recommends that an appropriate fraction of the funding for a lunar initiative be devoted to supporting fundamental scientific missions as they progress from small ground-based instruments, to modest orbital experiments, and finally, to the placement of facilities on the moon. The advanced technology should be tested by obtaining scientific results at each development stage.**

The committee urges the selection of a modest project for the early phase of the lunar program, such as a 1-m-class telescope for survey or pointed observations, that would provide useful scientific data and valuable experience for operating larger facilities in the future. The committee concluded that, in the long term, the chief advantage of the moon as a site for space astronomy is that it provides a large, solid foundation on which to build widely separated structures such as interferometers.

ASTRONOMY AND SOCIETY

Answering questions about the universe challenges astronomers, fascinates a broad national audience, and inspires young people to pursue careers in engineering, mathematics, and science.

- **The committee recommends enhancing astronomy's role in precollege science education by increasing the educational role of the national observatories, by expanding summer programs for science teachers, and by setting up a national Astronomy Fellowship program to select promising high school students for summer internships at major observatories.**

Astronomical research assists the nation, directly and indirectly, in achieving societal goals. For example, studies of the sun, the planets, and the stars have led to experimental techniques for the investigation of the earth's environment and to a broader perspective from which to consider terrestrial environmental concerns such as ozone depletion and the greenhouse effect.

Research in astronomy derives its support from the curiosity of human beings about the universe in which we live, from the stimulus it provides to young people to study science, from the synergistic benefits to other sciences, and from the unforeseen practical applications that occasionally ensue. While participating in the thrill of the discovery of new things "out there," society passes on something of value to future generations.

1

Recommendations

INTRODUCTION

Our Place in the Universe

Astronomy and astrophysics address questions about the origin and evolution of the planets, the stars, and the universe. In this century we have learned that the climates and weather patterns of planets in the solar system are driven by many of the same physical processes that create the earth's environment; that stars form out of clouds of gas and eventually die either in quiet solitude or spectacular explosions; that most of the common chemical elements are created in explosions of stars; that stars group together in isolated galaxies; that galaxies and clusters of galaxies stretch in sheets and filaments as far as the largest telescopes can see; and that the universe itself was born in a violent explosion some 15 billion years ago. Most amazingly, we have learned that the laws of nature that humans have discovered on the earth apply without modification to the farthest reaches of the observable universe.

Yet each new answer leads to new puzzles. What kinds of planets form around other stars? What triggers the formation of stars in our own galaxy and in other galaxies? What powers the enormous bursts of energy seen in some galaxies? How did galaxies themselves arise in the primitive universe? Where can black holes be found, and what are their properties? What is the ultimate fate of the universe? These are a few representative questions that capture the imaginations of astronomers and the general public and that stimulate young people to study mathematics, science, and engineering.

Discoveries of the 1980s

Observations with underground, ground-based, airborne, and orbiting telescopes during the 1980s produced important discoveries that advanced our knowledge in many areas of astronomy. The following is a selection of some of the more important advances and consolidations.

- The theory of the origin of the elements in the "Big Bang" received support from both astronomical observations of stars and sensitive experiments in particle physics.
- An orbiting satellite launched in 1989 began observing the relict radiation from the earliest years of the universe. Preliminary results indicate the need to revise existing theories of the formation of galaxies and clusters of galaxies.
- Evidence gathered shows that the radiation from as much as 90 percent of the matter of the universe has so far gone undetected.
- Quasars were found at extremely large distances and must have been formed when the universe was less than 10 percent of its present age.
- Einstein's prediction that the gravitation of matter could bend rays of light found application in the discovery that galaxies can act as lenses, refracting the light from more distant quasars.
- Surveys of large numbers of galaxies revealed that the universe is organized on larger scales than predicted by many cosmological theories.
- Increasing evidence suggested the possibility of giant black holes in the centers of some galaxies and quasars.
- An orbiting satellite surveyed the sky at infrared wavelengths and discovered disks of solid material, possibly the remnants of planet formation, orbiting nearby stars. It also found ultraluminous galaxies emitting 100 times as much energy in the infrared as at visible wavelengths.
- Supernova 1987A burst into prominence in our closest neighbor galaxy, the Large Magellanic Cloud. Subatomic particles called neutrinos from the supernova were detected in underground observatories, confirming theories about the death of stars and the production of the heavy elements crucial to life on the earth.
- Neutron stars spinning at nearly 1,000 revolutions per second were discovered by their regular pulses of radio radiation. Signals from these objects may constitute the most stable clocks in the universe, more accurate than any made by humans, and can be used to search for gravitational waves and as probes of the dynamics of star clusters.
- A deep probe of the interior of a star—our own sun—was achieved through a technique analogous to terrestrial seismology, measuring pressure

waves on the solar surface. These measurements established the extent of the solar convective zone and the dependence of rotation speed on depth in the sun.

- Experiments done with solar neutrinos hinted at new physics not included in standard textbooks.
- The mass and radius of Pluto were determined from observations of its satellite, Charon. Other studies of Pluto revealed the surprising fact that this small, cold planet has an atmosphere.
- Deuterium was discovered in the Martian atmosphere, and this isotope was used to measure the loss of water from Mars in the past.

The 1990s: The Decade of Discovery

The 1990s promise to be a decade of discovery. The first 10-m telescope, the Keck telescope in Hawaii, will come into operation early in the decade. This telescope and the others to follow will be the first very large optical and infrared telescopes constructed in this country since the epoch-making installation of the Hale 5-m telescope on Palomar Mountain over 40 years ago. The technological revolution in detectors at infrared wavelengths will increase the power of telescopes by factors of thousands. New radio telescopes will reveal previously invisible details at millimeter and submillimeter wavelengths. A technique called interferometry will combine optical or infrared light from different telescopes separated by hundreds of meters to make images thousands of times sharper than can be achieved with a single telescope. The four Great Observatories of the National Aeronautics and Space Administration (NASA) will view the cosmos across the infrared, visible, x-ray, ultraviolet, and gamma-ray portions of the electromagnetic spectrum. These instruments, orbiting above the earth's distorting atmosphere, will answer critical questions and may reveal objects not yet imagined.

PURPOSE AND SCOPE OF THIS STUDY

Charge to the Committee

The charge to the committee was as follows:

> The committee will survey the field of space- and ground-based astronomy and astrophysics, recommending priorities for the most important new initiatives of the decade 1990–2000. The principal goal of the study will be an assessment of proposed activities in astronomy and astrophysics and the preparation of a concise report addressed to the agencies supporting the field, the Congressional committees with jurisdiction over these agencies, and the scientific community. The study will restrict its scope to experimental and theoretical aspects of subfields involving remote observation from the earth and earth orbit, and analysis of astronomical objects; earth and planetary

> sampling missions have been treated by other National Research Council and Academy reports. Attention will be given to effective implementation of proposed and existing programs, to the organizational infrastructure and the human aspects of the field involving demography and training, as well as to suggesting promising areas for the development of new technologies. A brief review of the initiatives of other nations will be given together with a discussion of the possibilities of joint ventures and other forms of international cooperation. Prospects for combining resources—private, state, federal, and international—to build the strongest program possible for U.S. astronomy will be explored. Recommendations for new initiatives will be presented in priority order within different categories. The committee will consult widely within the astronomical and astrophysical community and make a concerted effort to disseminate its recommendations promptly and effectively.

The committee agreed that the primary criterion determining the order of priorities would be the committee's best estimate of the scientific importance of each initiative. In forming its judgment of scientific importance, the committee also took into account cost-effectiveness, technological readiness, educational impact, and the relation of each project to existing or proposed initiatives in the United States and in other countries.

In a letter to the committee commenting on the initial charge, NASA's associate administrator for space science pointed out that NASA's solar physics research program contains investigations of the sun viewed both as a star and as a power source for the solar system, and that many of NASA's solar physics missions have a strong coupling to in situ measurements, which lie outside the purview of this committee. NASA requested that, for these reasons, solar physics space missions not be prioritized together with purely astronomical missions. The committee concurred with this request, since it reflected the nature of the subject and of the funding sources, but considered that ground-based solar astronomy remained within its charge. Independently, the Solar Astronomy Panel established by the committee [see the *Working Papers* (NRC, 1991) of this report] elected to develop an integrated plan for solar research incorporating both ground- and space-based initiatives.

The committee surveyed the entire field of astronomy and astrophysics as defined by its charge and attempted to engage everyone in U.S. astronomy who had an interest in being heard. More than 300 astronomers, listed in Appendix C, served on the 15 panels whose separately published reports (*Working Papers*) contain important advisory material that was considered by this committee. An additional 600 or so astronomers contributed directly to this report by their letters, essays, or oral presentations at open meetings; more than 15 percent of all U.S. professional astronomers played an active role in some aspect of this report. Distinguished colleagues from throughout the world contributed valuable essays and letters. The committee also profited from discussions with dedicated people in Congress and on congressional staffs, and with personnel in the funding agencies, in the Office of Management and Budget, and in other executive offices.

In carrying out its charge, the committee describes prioritized equipment initiatives that reflect its best judgment about what facilities will most advance the central goal of astronomy: understanding the universe we live in. However, the committee recognizes that there can be no research without researchers, teaching without students, or observational progress without advances in technology. An infrastructure of students, researchers, and equipment and a vigorous program of theoretical research must exist to support new work, or the new initiatives will not succeed. This committee therefore prefaces its discussion of new initiatives with (1) recommendations for strengthening the infrastructure for ground-based astronomy and (2) a discussion of the need for a balanced strategy for space astrophysics.

Contents of This Report

This report presents a prioritized program for the 1990s that balances the development of new facilities with support for existing facilities and for the research of individual scientists. The present chapter, Chapter 1, summarizes the prioritized recommendations for new instrumental initiatives. Other recommendations appear in the context of specific discussions in the chapters on existing programs, on computing, on the lunar initiative, and on policy opportunities.

Chapter 2 describes some of the scientific opportunities of the next decade and Chapter 3, some of the most important ongoing programs. Chapter 4 presents a more detailed scientific and technical justification for the recommended new initiatives. Chapter 5 outlines the influence of the computer revolution on astronomy. Chapter 6 evaluates the potential role of observatories on the moon in the nation's Space Exploration Initiative. Chapter 7 discusses some important policy issues in ground- and space-based astronomy. Finally, Chapter 8 highlights some of the ways astronomy benefits the United States and the world. Appendix A defines some of the most common and important astronomical terms used in this report. Appendix B gives some basic statistics on the current demography and funding of astronomical research. Appendix C lists the scientists who served on the panels established by the committee to help carry out this decennial survey.

RECOMMENDATIONS FOR STRENGTHENING GROUND-BASED INFRASTRUCTURE

- **The highest priority of the survey committee for ground-based astronomy is the strengthening of the infrastructure for research, that is, increased support for individual research grants**

and for the maintenance and refurbishment of existing frontier equipment at the national observatories.

By any quantitative measure, the research infrastructure has deteriorated seriously in the last decade: support for maintenance and refurbishment of facilities and for individual research grants in astronomy and astrophysics has declined as a fraction of the total budget of the National Science Foundation (NSF), as a fraction of the NSF's total astronomy budget, on a per-astronomer basis, and on the basis of real-dollar expenditures. NSF funding for astronomy has decreased for nearly a decade despite an explosion in research discoveries, a major expansion in the number and complexity of observational facilities, and a large increase in the number of practicing astronomers (see Appendix B). The consequences of this decline include the loss of key technical personnel, limitations on young scientists' participation in the research program, the delay of critical maintenance, the inability to replace old and obsolete equipment, and a lack of funds to pay for scientists to travel to observatories or to reduce data. The situation has reached critical dimensions and now poses a threat to the continued success of U.S. astronomy.

- **The committee recommends that the NSF increase its support for annual operations, instrument upgrades, and maintenance of national research facilities to an adequate and stable fraction of their capital cost. The NSF should include appropriate financial provision for the operation of any new telescope in the plan for that facility. The committee estimates that appropriate remedial actions will require increasing the operations, maintenance, and refurbishment budgets for the observatories now in existence by a total of $15 million per year.**

The recommended annual increase will serve to repair the effects of deferred maintenance at the National Optical Astronomy Observatories (NOAO) and the National Radio Astronomy Observatory (NRAO), upgrade receivers and correlators at NRAO to improve the performance of the Very Large Array (VLA) by a factor of 10, provide needed computational resources to deal with large-format arrays of optical and infrared detectors at NOAO, replace antiquated equipment at the National Astronomy and Ionosphere Center (NAIC), and hire new technical staff to service millimeter receivers, infrared arrays, and advanced optics. If maintained for a decade, the increases will restore the infrastructure to a healthy working condition.

- **The committee recommends that individual research grants be increased to an adequate and stable fraction of the NSF's total operations budget for astronomy. In order to gather and analyze the large amounts of data that will become available with new instrumentation, to allow young researchers to take advantage of**

the new opportunities for discovery, and to restore support for theoretical astrophysics, the individual grants budget should be increased by $10 million per year.

The grants program supports the activities of graduate and postdoctoral fellows who become the trained scientists of the future; it supports the design of new instruments, the purchase of computers and the analysis of data, and the development of theories to explain and motivate observations and to predict new results. Many of the fundamental discoveries of the 1980s were made using ground-based facilities and with support provided by individual research grants. At present, only one grant in two is able to support a graduate student, and only one grant in four can support a postdoctoral fellow.

The recommended annual increase for the grants program is meant to accomplish three purposes: first, improve the chances for first-time applicants, many of whom are young scientists with no other means of research support, to receive grants ($1.5 million per year); second, increase the average grant size by $20,000 to a total of $80,000, which is approximately the size needed to support and train an individual postdoctoral fellow ($7.5 million per year); and third, as discussed below, help establish an astrophysics theory program within the NSF ($1 million per year).

- **Within the astronomy grants program of the NSF, the committee recommends that theoretical astrophysics be given additional visibility and resources.**

Theory provides the basic paradigms within which many observations are planned, analyzed, and understood. As discussed in Chapter 3, adequate support for theory is necessary in order to realize the full benefits from existing and recommended observing facilities.

ACHIEVING A BALANCED SPACE PROGRAM

Overall Strategy

For space-based astrophysics, the most urgent need is to continue to develop a program that balances the enormous power of the largest observatories with smaller missions of reduced complexity and shorter development times, while maintaining a healthy research and analysis infrastructure. Large missions are essential for solving certain fundamental scientific problems, but their development can require as long as two decades. Moderate and small missions can respond more rapidly to changing scientific priorities and to new technical or instrumental breakthroughs. Rapid access to space attracts talented young instrumentalists and stimulates innovation. The committee believes that a greater involvement of scientists in engineering and management issues could improve the efficiency and scientific return of space astronomy missions of all sizes. In order to achieve its scientific goals, NASA must carry out its plans to devote adequate resources to the Research and Analysis program and to Mission Operations and Data Analysis. This support is crucial for analyzing and interpreting data, for understanding the implications of the data through theory, and for developing new technologies for future space missions. The funding requirements for new instrumental initiatives within a balanced space program are listed in Table 1.1.

Before deciding on its final recommendations for new equipment initiatives, the committee considered the implications of the recently discovered problems with the Hubble Space Telescope.

Significance of Large Space Observatories

The operation of the first Great Observatory, the Hubble Space Telescope (HST), began disappointingly. Initial observations revealed that a flaw in its primary mirror would prevent HST from achieving its design resolution and sensitivity without corrective action. A failure in the testing and quality control of known technology apparently caused the flaw in the mirror. Investigatory committees have been appointed to identify precisely what happened, why it happened, and how to protect future programs from other major mistakes. In Chapter 7, the committee describes its view that a strong involvement by scientists is critical to the success of space projects of all sizes.

This committee has reexamined the justification for large-scale space astronomy programs, taking into account both the failure to meet specifications in the HST program and NASA's record of successes in carrying out other complex missions at the frontiers of science and technology. As will be clear from discussions later in this chapter and in Chapters 3 and 4, the committee has concluded that in some cases only large-scale programs can answer some of the most fundamental astronomical questions. The four Great Observatories currently planned by NASA for the 1990s cover a large fraction of the electromagnetic spectrum with the sensitivity and resolution required to make progress on frontier problems in astrophysics. The Great Observatories are the HST, the Gamma Ray Observatory (GRO), the Advanced X-ray Astrophysics Facility (AXAF), and the Space Infrared Telescope Facility (SIRTF).

The Great Observatories embody the ideal of "small science" made possible by large facilities; they allow individual investigators or small groups of investigators to carry out frontier research programs. The typical number of researchers is only four per proposal for the approved HST programs. Investigators at about 200 different institutions were awarded observing time for the first year of HST operations.

Nevertheless, the committee believes that a balanced space astronomy program would put increased emphasis on more frequent and less costly missions. Modest, cost-efficient missions can respond more rapidly to changing scientific ideas, to technological developments, and to the need for training young researchers. Thus the committee calls for a greater frequency of astrophysics Explorer missions and limits the recommendations for new, large space programs to one mission, SIRTF. SIRTF has superb science potential, is technologically well developed, and has been preceded by two smaller and successful precursor missions.

RECOMMENDED NEW EQUIPMENT INITIATIVES

Ground and Space Initiatives

- **The Astronomy and Astrophysics Survey Committee recommends the approval and funding of the set of new equipment initiatives listed in Table 1.1.**

Table 1.1 lists the recommended hardware initiatives that fall within the committee's charge, along with the committee's best estimates of their costs (in 1990 dollars). Detailed descriptions of the principal programs appear in Chapter 4. The programs are arranged in large, moderate, and small categories, depending on the scale of resources required. Large and moderate programs are listed in order of scientific priority. Since one set of agencies (NSF, DOE, and DOD) supports ground-based programs and a different agency (NASA) supports space-based programs, the committee has separated the new initiatives into ground- and space-based projects. Two areas to which the committee assigns high priority in other parts of this report—science education (Chapter 7) and theoretical astrophysics (Chapter 3)—do not appear explicitly in Table 1.1 because most of their support is provided from grants to individual researchers and not through facilities.

The costs presented in Table 1.1 have various origins. Certain programs such as SIRTF, the two 8-m ground-based telescopes, and the Millimeter Array (MMA) have already benefited from extensive design studies; their costs are well defined and should not change appreciably except as a result of inflation or delays in implementation. For SIRTF and other space missions, the costs include construction of the entire facility and operation through launch plus 30 days. For the large ground-based telescopes, the costs include a first complement of instruments and rudimentary computer analysis. Costs for the other recommended programs are based on discussions with agency personnel or on material presented to the panels [see the *Working Papers* (NRC, 1991)]. In most cases, the programs have been studied in enough detail under the auspices of the funding agencies that the costs are well approximated by the values listed. In arriving at the prioritized list, the committee made its own informal assessment of the realism of the proposed costs, taking into account technological readiness and the heritage from previous projects.

TABLE 1.1 Recommended Equipment Initiatives and Estimated Costs

Ground-based	Decade Cost ($M)	Space-based	Decade Cost ($M)
Large Programs			
Infrared-optimized 8-m telescope	80	Space Infrared Telescope Facility (SIRTF)	1,300
Millimeter Array (MMA)	115		
Southern 8-m telescope	55		
Subtotal ground-based	250	Subtotal space-based	1,300
Moderate Programs			
Adaptive optics	35	Dedicated spacecraft for FUSE	70
Optical and infrared interferometers	45	Stratospheric Observatory for Far-Infrared Astronomy (SOFIA)	230
Several shared 4-m telescopes	30	Delta-class Explorer acceleration[a]	400
Cosmic-ray telescope (Fly's Eye)	15	Astrometric Interferometry Mission (AIM)	250
Large Earth-based Solar Telescope (LEST)	15	International collaborations on space instruments	100
VLA extension	32		
Subtotal ground-based	172	Subtotal space-based	1,050
Illustrative Small Programs[b]			
Two-micron survey	5	*Small Explorer acceleration*	100
Infrared instruments	10	Orbiting planetary telescope	50
Cosmic background imager	7	VSOP/RadioAstron	10
Laboratory astrophysics	10	Laboratory astrophysics	20
National astrometric facility	10		
300-m antenna in Brazil	10		
Stellar oscillations instrument	3		
Optical surveys	6		
Neutrino supernova watch	10		
Subtotal ground-based	71	Subtotal space-based	180
Total ground-based	493	Total space-based	2,530
DECADE TOTAL			3,023

[a] Examples include gamma-ray spectroscopy, submillimeter spectroscopy, and x-ray imaging.
[b] Three small ground-based programs and one space initiative are highlighted by italics since they were regarded by the committee as being of special scientific importance.

In addition to the instrument initiatives recommended in this section, the committee stresses that progress in astronomy requires a vigorous program of theoretical research and laboratory astrophysics, in addition to a balanced program of observation. The vital role of theory and that of laboratory astrophysics are discussed further in Chapter 3.

This report's predecessor, *Astronomy and Astrophysics for the 1980's* (NRC, 1982), widely known as the "Field Report," was well received in large part because astronomers made for themselves the difficult priority choices. The two highest-ranking major programs (out of a total of four recommended new programs) were funded; the two highest-ranking moderate programs, plus two other programs, out of a total of seven recommended new programs, were initiated; and the highest-ranking small program, plus one other, received funding. The completion of two of these initiatives [AXAF and the Far Ultraviolet Spectroscopy Explorer (FUSE) space observatories] remains a goal of this committee. The total estimated budget for the Field Committee's recommended new programs was $1.7 billion (1980), or approximately $2.6 billion in 1990 dollars, compared to an estimated $3.0 billion (1990) for the items listed in Table 1.1.

The Combined Equipment List

The committee presents in order of priority in Table 1.2 its combined list of new equipment initiatives, independent of agency and independent of the location of the facility, ground or space. The major individual items are described briefly below and more fully in Chapter 4.

Tables 1.1 and 1.2 are prioritized lists for new hardware initiatives. The committee emphasizes again that other areas also have high priority. The restoration of the infrastructure, including the refurbishment of existing equipment and increased support for individual research grants, is the highest-priority recommendation for ground-based astronomy, as discussed above. The need to establish a program with the proper balance between large, moderate, and small missions is the highest priority for space astrophysics. The committee stresses that progress in astronomy requires a vigorous program of theoretical research and laboratory astrophysics, as well as a balanced program of observation. The vital roles of theory and of laboratory astrophysics are discussed further in Chapter 3. In addition, astronomy has a long tradition of support by state, private, and even international sources. The committee did not attempt to prioritize initiatives supported by nonfederal sources of funding, but the contribution of such projects is important and is included in Chapter 3, a discussion of ongoing and planned programs.

Small Projects and Technological Initiatives

The committee decided to maintain separate lists for ground- and space-based small projects and for technology initiatives in recognition of the ability of funding agencies to respond quickly to new scientific opportunities and to advances in instrumentation. Table 1.1 lists some illustrative small projects that have great scientific merit.

TABLE 1.2 Recommended Equipment Initiatives (Combined Ground and Space) and Estimated Costs

Initiative	Decade Cost ($M)
Large Programs	
Space Infrared Telescope Facility (SIRTF)	1,300
Infrared-optimized 8-m telescope	80
Millimeter Array (MMA)	115
Southern 8-m telescope	55
Subtotal for large programs	1,550
Moderate Programs	
Adaptive optics	35
Dedicated spacecraft for FUSE	70
Stratospheric Observatory for Far-Infrared Astronomy (SOFIA)	230
Delta-class Explorer acceleration	400
Optical and infrared interferometers	45
Several shared 4-m telescopes	30
Astrometric Interferometry Mission (AIM)	250
Cosmic-ray telescope (Fly's Eye)	15
Large Earth-based Solar Telescope (LEST)	15
VLA extension	32
International collaborations on space instruments	100
Subtotal for moderate programs	1,222
Subtotal for illustrative small programs	251
DECADE TOTAL	3,023

Table 1.3 lists, but not in any prioritized order, the committee's assessment of the most important technological initiatives for the 1990s that will form the basis for frontier science in the decade 2000–2010. The costs given in Table 1.3 represent this committee's best estimate of what is required to make major technological progress in each area. Details of the proposed technology development programs appear in the *Working Papers* (NRC, 1991).

Explanation of New Equipment Initiatives

This section provides thumbnail sketches of the large and moderate programs recommended by the committee. More complete technical and scientific descriptions appear in Chapter 4.

TABLE 1.3 Technology Development for the 1990s and Estimated Costs

Technology Initiative	Decade Cost ($M)
Ground-based	
Infrared detector arrays	25
Optical detector arrays	10
Solar neutrino experiments	15
Dark matter detectors	10
Digital archive	15
Gamma-ray airshower detectors	5
Radio technology	15
Subtotal for ground-based technology	95
Space-based	
Optical and infrared interferometry in space	50
Technology for next-generation observatories:	
Large space telescope technology	50
Submillimeter receiver and telescope technology	75
High-energy mirror and detector technology	50
Subtotal for space-based technology	225
DECADE GRAND TOTAL	320

LARGE PROGRAMS

Ground-based Astronomy

Three programs have the greatest importance for ground-based astronomy. They are discussed in the committee's order of scientific priority, which corresponds also to the most appropriate temporal sequence for their implementation.

Infrared-Optimized 8-m Telescope. The highest-priority recommendation for a large ground-based facility is for the construction of an 8-m-diameter telescope on Mauna Kea, Hawaii; this telescope and its instrumentation should be optimized for low-background, diffraction-limited operation at infrared wavelengths from 2 to 10 μm. The planned telescope will be a unique facility, using revolutionary infrared array detectors to make high-spatial- and high-spectral-resolution observations of objects as various as volcanoes on the satellite Io, the inner parts of protoplanetary disks around nearby young stars, and distant galaxies. The telescope builders will benefit from the past decade of intense development in the technology for casting and polishing large mirrors. Chapter 4 discusses the complementarity of the major infrared and submillimeter facilities recommended in this report.

Millimeter Array. The proposed Millimeter Array will be an imaging telescope operating at millimeter radio wavelengths with high spatial and spectral resolution. The array will consist of 40 transportable antennas, each 8 m in diameter. Scientific projects advanced by the MMA will range from extragalactic programs to solar and planetary programs and include studying flares on the sun, imaging the outer parts of protoplanetary disks, and probing motions in the cores of distant, infrared-luminous galaxies. The MMA builds on techniques pioneered by the VLA, which operates at centimeter wavelengths, and by two university observatories that operate small arrays at millimeter wavelengths.

Southern 8-m Telescope. The proposed Southern Hemisphere 8-m telescope will be a twin of the Northern Hemisphere infrared-optimized 8-m telescope, except that the southern telescope and its instrumentation will be optimized for optical and near-ultraviolet wavelengths. The southern 8-m telescope will provide U.S. astronomers with a vital window through which to view objects that are best observed from the Southern Hemisphere, such as the center of our galaxy and our closest neighboring galaxies, the Magellanic Clouds.

These three large ground-based initiatives will provide state-of-the-art research facilities for hundreds of researchers working at dozens of different institutions. For example, the 4-m telescopes, the largest instruments at NOAO, typically have 150 users each per year (20 percent of whom are students), with an average of about 1.5 astronomers per approved proposal. The VLA radio telescope has about 600 users per year, with an average of about 3 astronomers per research program.

Space-based Astronomy

SIRTF. The highest priority for a major new program in space-based astronomy is the Space Infrared Telescope Facility, a 0.9-m cooled telescope in a spacecraft to be launched by a Titan IV-Centaur into high earth orbit. SIRTF will operate as a national facility, with more than 85 percent of the observing time during its five-year lifetime available to individuals and small groups of investigators from the general astronomical community. Across the wavelength region from 3 to 200 μm, SIRTF will be up to a thousand times more sensitive than other space- or ground-based telescopes, and over a million times faster than other instruments for surveying, mapping, or obtaining spectra across large, complex regions (see Chapter 4). The technical heritage of SIRTF includes two infrared telescopes launched and successfully operated as part of NASA's Explorer program.

MODERATE PROGRAMS

Ground-based Projects

Adaptive Optics and Interferometry. The two highest-priority moderate programs for ground-based astronomy involve the application of new techniques for optical and infrared observing. The highest priority is to apply technologies collectively called adaptive optics to overcome the blurring effects of the earth's atmosphere on time scales of a few hundredths of a second and thereby enhance the sensitivity and spatial resolution of new or modernized telescopes. The second highest priority is to use techniques of interferometry, previously used successfully by radio astronomers, to join separated infrared or optical telescopes. A linked group of telescopes would have a resolving power equivalent to that of a single instrument as large as the distance separating the individual telescopes. Demonstration projects have proven successful for both of these techniques, suggesting that adaptive optics should be applied to a broad range of telescopes and that several interferometers, more powerful than existing ones, should be supported.

4-m Telescopes. While discovery often follows the opening of new frontiers in sensitivity, wavelength, or angular resolution, detailed understanding usually requires detective work by large numbers of scientists with access to appropriate investigative tools. The basic "detective tool" for standard investigations in the 1990s will be a 4-m-class optical or infrared telescope. Federal funds should be used in combination with state and private funds when possible to construct several new 4-m telescopes. Advances in technology will make it possible to build and operate these facilities more economically than the first generation of 4-m telescopes at NOAO, endorsed by the "Whitford Report" (NRC, 1964). University involvement in operation and management of these facilities will provide opportunities for students to perform exploratory or long-term programs and to help develop instrumentation.

Fly's Eye. Cosmic rays consist primarily of protons and the nuclei of heavy atoms. The existing Fly's Eye telescope has detected the fluorescent trails of over 200 cosmic rays more energetic than 10^{19} eV. At present, no one knows how particles can be accelerated to such high energies, what their composition is, or whether such particles originate inside or outside of the galaxy. A new Fly's Eye telescope, with a factor-of-10 improvement in sensitivity and better spatial resolution, would help determine the anisotropy, the energy spectrum, and the composition of cosmic rays in the energy range 10^{19} to 10^{20} eV.

Large Earth-based Solar Telescope. U.S. solar astronomers have entered into an international collaboration with scientists from eight other countries to build the world's best ground-based solar telescope, to be located in the Canary Islands. The Large Earth-based Solar Telescope (LEST) is a 2.4-m telescope that will obtain diffraction-limited images and spectra of the sun using the techniques of adaptive optics. The United States will contribute the most technically challenging part of the telescope, the adaptive optics system.

VLA Extension. The Very Large Array of the NRAO has proven extraordinarily productive in studying objects as diverse as comets, planets, the sun, other stars, interstellar clouds, and distant radio galaxies and quasars. The arc-second imaging of the VLA will complement the thousandth-of-an-arcsecond capability of the Very Long Baseline Array (VLBA) when the latter is completed in 1992. The VLA extension will add four new telescopes and ancillary hardware to bridge the lack of baselines between telescopes needed for angular resolutions between a tenth and a thousandth of an arcsecond in the performance of the VLA and the VLBA.

Space-based Projects

Dedicated Spacecraft for FUSE. The committee recommends that NASA augment the Explorer program sufficiently to convert the Far Ultraviolet Spectroscopy Explorer into a Delta-launched experiment using its own dedicated spacecraft. In 1989, NASA selected FUSE for development in the Explorer program. This committee, like the Field Committee a decade ago, believes FUSE will produce extraordinary scientific results. The committee is concerned that continued linking of FUSE to the Shuttle program through a reusable Explorer spacecraft will unnecessarily delay the mission and increase its cost. A dedicated spacecraft will enhance the scientific potential of the FUSE mission, providing an optimized orbit, simpler operations, and a longer mission lifetime.

SOFIA. The proposed Stratospheric Observatory for Far-Infrared Astronomy (SOFIA) is a 2.5-m telescope mounted in a Boeing 747 aircraft and optimized to study infrared and submillimeter wavelengths from above most of the earth's water vapor. SOFIA provides the highest-resolution spectroscopy of any planned facility for wavelengths between 30 and 300 microns, and its large aperture will yield high-spatial-resolution observations that will complement SIRTF's capabilities. In addition to its scientific value, SOFIA will be an excellent facility for the training of the next generation of experimentalists and for the rapid development of advanced instrumentation. The predecessor to SOFIA is the Kuiper Airborne Observatory (KAO), which has trained some 40 PhD astronomers in an environment that subjects both students and instrumentation to many of the rigors of space missions. SOFIA represents the natural evolutionary replacement for the KAO and is a joint project with Germany.

Explorers. The committee believes that flying six Delta-class Explorer missions for astrophysics in the next 10 years will play a key role in revitalizing space astronomy. Each of the panels dealing with space observations [*Working Papers* (NRC, 1991)] identified examples of forefront science that could be carried out with such experiments. The missions should be chosen through the peer review process, with specific emphasis on modest scope, rapid execution, and careful cost control. The active involvement of the selected science team in all phases of the program is critical to the success of these missions.

Astrometric Interferometry Mission. Astrometry, which is concerned with the measurement of the positions of celestial sources, ranks among the oldest and most fundamental branches of astronomy and now lies on the verge of a technological revolution. The application of interferometric techniques in space with telescope separations of a few hundred meters may enable a 1,000-fold improvement in our ability to measure positions. An Astrometric Interferometry Mission (AIM) with 3- to 30-millionths-of-an-arcsecond accuracy could detect Jupiter-sized planets around hundreds of stars up to 500 light-years away.

International Collaborations. The ability to place U.S. instruments on foreign spacecraft is a cost-effective way to provide access to space. NASA has often taken support for such experiments out of the strained Explorer budget. A budget line for international collaborations will allow NASA to undertake more of these advantageous joint ventures.

SMALL PROGRAMS

Small programs can be carried out relatively quickly in response to new scientific or technological developments, focusing research into the currently most profitable areas and making possible greater participation by young astronomers. Some of the most exciting scientific results of the past decade have come from modest, cost-effective programs.

- **The committee recommends that an increased emphasis be given in the astronomy research budget to small and moderate programs.**

However, it would be counterproductive to set detailed priorities for small programs for an entire decade, because this would interfere with the desired flexibility and rapid response. The federal agencies can peer review small programs on a timely basis. The committee therefore lists illustrative examples of small programs that are of high priority at the present time. Over the next several years some of the programs listed here will likely be either funded or replaced by other small initiatives.

Highlighted by italics in Table 1.1 are three small ground-based programs and one small program for space that the committee regards as being particularly important at this time. Other small programs of high quality are discussed in the *Working Papers*; the committee believes that many other outstanding small programs will be identified throughout the 1990s.

Ground-based Projects

Two-Micron Survey. The only survey of the sky in the near infrared ($\sim 2\ \mu m$) occurred over 20 years ago. Today, a pair of 1-m telescopes, one for each hemisphere, equipped with modern near-infrared array detectors, can completely survey the sky at three wavelengths in less than two years, reaching a level 50,000 times fainter than the earlier survey.

Infrared Instrumentation. The ongoing revolution in infrared technology can improve the data-collection capability of existing telescopes by factors of tens of thousands by replacing single-element detectors with large, multielement arrays. Building instruments that incorporate arrays will enhance both imaging and spectroscopic capabilities.

Cosmic Background Imager. On angular scales smaller than a few degrees, the cosmic background radiation reflects conditions in the early universe at an age of only 100,000 years. Recent technological advances suggest that a Cosmic Background Imager could reach the levels of sensitivity required to search for variations in the brightness of the background in different directions. Primordial density fluctuations that may be the precursors of galaxies can be revealed by these variations in brightness.

Laboratory Astrophysics. The interpretation of observations from ground-based telescopes often depends on the results of laboratory experiments concerning basic atomic, molecular, or nuclear data. In some cases, theoretical calculations are necessary because the appropriate quantities cannot be measured. Since many important experiments in laboratory astrophysics are primarily of interest to astronomers, some laboratory research will require direct funding from astrophysics resources.

Other Programs. Other important small projects listed in Table 1.1 are a ground-based facility for making long-term astrometric measurements, U.S. participation in an international project to build a 300-m radio telescope in Brazil, instrumentation to study the "seismology" of stars, optical all-sky surveys of galaxies with modern electronic detectors, and systematic monitoring to detect neutrino bursts from supernovae. Details of these and other projects are described in the *Working Papers.*

Space-based Projects

Small Explorers. The committee highlights an acceleration of the Small Explorer (SMEX) program to be carried out within tight budgetary constraints with the goal of making possible five astronomy SMEX missions in the 1990s. These small new missions should be selected by the peer review process and launched on Scout-class rockets. SMEX missions will help train the future leaders of space astronomy and will provide a rapid method for executing certain well-defined, high-priority projects. Some of the panel reports in the *Working Papers* contain excellent ideas for SMEX payloads.

Other Projects. Other space projects that will return important data at relatively low cost include U.S. participation in a German orbiting planetary telescope, and orbiting very long baseline interferometry (VLBI) experiments being conducted with the Soviet Union and Japan. The success of future

NASA missions like AXAF, SIRTF, and SOFIA depends in part on improved knowledge of the physical and chemical properties of atoms, molecules, and dust grains that can be obtained only with a vigorous program of laboratory measurement. Laboratory astrophysics provides an essential key for the success of the NASA missions planned for the 1990s.

TECHNOLOGY DEVELOPMENT

The technology developed today is used in the science of tomorrow. The ground- and space-based technology programs include new methods for improving the performance of detectors, developing new types of telescopes, and making astronomical data widely available.

Ground-based Technology

The largest and most efficient detectors allow optimum use of the nation's investment in telescopes. Infrared detectors are improving rapidly with the aid of technology developed for national security applications. With continuing access to the fruits of this development, only a modest investment is required to optimize and purchase infrared detectors for astronomical applications and to improve relatively mature optical devices.

Promising new techniques are being developed to observe the solar neutrinos emitted during the key energy-producing reactions in the solar interior, including proton-proton collisions and the decay of ^{7}Be. These techniques can reveal fundamental aspects of how the sun shines and, at the same time, provide important information about particle physics. If technology development proves successful, these experiments could produce data in the second half of the 1990s.

Novel or improved detectors must be developed to detect ultrahigh-energy gamma rays from astronomical sources and to observe exotic particles that could constitute the "dark matter" of the universe.

The ground- and space-based observatories of the 1990s will produce immense amounts of data. As discussed in Chapter 5, the archiving of these data is a high scientific priority.

Developments in three areas have particularly great potential to enhance the efficiency of radio telescopes: low-noise receivers for millimeter and submillimeter wavelengths, broad-bandwidth recording systems and data links for VLBI, and focal plane arrays.

Space-based Technology

Infrared or optical interferometers in space many kilometers in size, either in orbit or on the moon, offer ultrahigh angular resolution. As discussed in Chapter 6, a phased technology development program in this area with intermediate technical and scientific milestones, including ground-based and modest orbiting experiments, is required to prepare the way for such major programs in the next century.

Astronomers will uncover new phenomena with the Great Observatories throughout much of the 1990s and into the next decade. If the past is any guide, however, the problems we solve will lead to deeper questions and new directions. We must begin now the conceptual planning and technological development for the next generation of astronomy missions to follow the Great Observatories.

One example of a next-generation space observatory is a large space telescope, a 6-m telescope that would combine the light-gathering power of a large ground-based telescope with the excellent image quality, ultraviolet capability, and low-infrared background that are achievable in space. Other possible missions with great scientific potential include a large x-ray telescope equipped with detectors capable of simultaneous imaging and spectroscopy; a submillimeter observatory consisting of a deployable 10-m telescope or an orbiting array of smaller telescopes operating as an interferometer; a single large radio telescope many kilometers across; and five orbiting radio telescopes of 100-m diameter forming an array that surrounds the earth.

- **The committee recommends that NASA pursue the technology initiatives listed in Table 1.3 as a prerequisite for initial definition studies of the next-generation space astronomy missions to be initiated early in the second half of the 1990s.**

The panel reports in the *Working Papers* (NRC, 1991) contain detailed discussions of technologies that require further study. The development of advanced projects should proceed in a step-by-step manner with frequent tests of technologies and the involvement of key personnel by means of scientific missions of increasing scope. These studies would provide the basis for the selection, by the turn of the century, of a new mission to follow the Great Observatories.

The scientific imperatives and the infrastructure available at the time of selection will influence which missions are chosen. Technical issues will include the construction and control of lightweight systems, the capabilities of launch vehicles, advances in robotic construction techniques, and the availability of facilities on the moon. The technology development programs listed in Table 1.3 will provide part of the factual basis required for decisions about future astronomical missions.

2

Science Opportunities

INTRODUCTION

Astronomers believe that the atoms of our bodies were created in the primordial explosion that marked the beginning of the universe and in the nuclear fires burning within stars. Some of these atoms, scattered throughout space, ultimately formed planets, soil, and organic molecules. Thus to understand our human beginnings we must understand the life history of stars and galaxies and even the whole universe. Whereas astronomers of the past were concerned more with charting the stars in a permanent cosmos, astronomers today study evolution and change.

Astronomical findings in this century have raised questions about life cycles in space. We have found that the Great Red Spot of Jupiter is not a fixed birthmark, but results from a continuing struggle of gases in motion. We have witnessed stars in the act of formation, still wrapped in the gas and the dust out of which they condensed. We have seen other stars exploding, having first spent their nuclear fuel and then collapsing under their own weight. And in the debris from stellar explosions, we have found oxygen and carbon and other elements essential to life. We have discovered huge streams of matter propelled from the centers of galaxies at nearly the speed of light. We have found that galaxies evolve. We have learned that galaxies are not scattered evenly in space, but are bunched together in filaments and sheets and other large groupings whose origins have not yet been explained. Finally, we have gathered evidence that the entire mass of the universe began in a state of fantastic compression, some 10 billion to 20 billion years ago.

How did the universe come into being in the first place? What determined its properties? Will it keep expanding forever or instead collapse on itself? A century ago, such questions were considered to lie outside the domain of science. Today, they are central to the field of astronomy.

Progress in astronomy is driven by advances in technology. In the 1930s, new communication devices led to the reception of radio waves from space. For thousands of years before, visible light had been our only way of seeing the universe. Since the 1950s, rockets and satellites have recorded infrared radiation, ultraviolet radiation, and x-rays emitted from space. Such radiation, invisible to the human eye, has revealed completely new features of many astronomical objects and has announced some objects not before known. New electronic detectors have replaced photographic plates, resulting in a 100-fold increase in sensitivity and a broader range of available wavelengths. High-speed computers have revolutionized theoretical astronomy by permitting the simulation of millions of interacting stars or galaxies. Electronically linked to combine the data from different antennas, large arrays of radio telescopes can work together as if they were one giant eye.

As discussed in Chapters 1 and 4, astronomical exploration in the 1990s will take advantage of novel technologies to make new instruments of startling power. In equipping ourselves for the future, diversity must accompany precision. We theorize and forecast as well as we can, but if the past is a guide, some of the next decade's discoveries will catch us off guard. In astronomy, the frontiers surround us.

OUR SOLAR SYSTEM AND THE SEARCH FOR OTHER PLANETS

The Formation and Evolution of Our Solar System

In the middle of the 18th century, the German philosopher Immanuel Kant proposed that our system of planets and sun condensed out of a great rotating cloud of gas and dust. This proposal, called the nebular hypothesis, has gained enormous observational and theoretical support and is favored today. Kant suggested that a primitive gaseous cloud slowly contracted under the inward pull of its own gravity. The central, densest regions formed the sun. The outer regions collapsed along the axis of rotation, because of gravitational forces, but did not fall directly toward the nascent sun, because of centrifugal forces pushing outward. Caught between these two forces, the material formed a flattened shape, called a protostellar or protoplanetary disk, in orbit about the sun.

In processes still under investigation by astronomers using computer simulations, material in the disk formed into dense condensations that eventually coalesced into planets—some small and rocky, like Mercury, Venus, Earth, and Mars, and others further from the sun and composed primarily of hydrogen gas, like Jupiter, Saturn, Uranus, and Neptune. Almost all the remaining material in the disk was swept out of the solar system, except for trace amounts of dust and small bodies like comets and asteroids.

Two different phases in this evolutionary scenario received observational support in the 1980s. Astronomers working with data from an orbiting telescope in space, the Infrared Astronomical Satellite (IRAS), found young stars forming in the nearby constellation of Taurus that appeared to be surrounded by disks roughly the size of a nascent solar system. But IRAS alone lacked the ability to discern the small structures and small velocities necessary to make a conclusive judgment. Astronomers using arrays of millimeter wavelength telescopes linked together to act as a single large telescope demonstrated that the material around the young stars was in orbit around the stars and was present in a quantity sufficient to make solar systems. IRAS found supporting evidence for a later phase in the life history of solar systems when it discovered that as many as one-quarter of all nearby stars are surrounded by disks of orbiting particles that may be the debris left over from the formation of planets (Plate 2.1).

Understanding how stars and planets form will be one of the major scientific themes of astronomy in the 1990s. Progress will come from increases in sensitivity and spatial resolution at radio, infrared, and optical wavelengths. The Space Infrared Telescope Facility (SIRTF) will be a thousand to a million times more sensitive than its predecessor IRAS, enabling scientists to search for and characterize protoplanetary disks at all stages of evolution. Observations with the Millimeter Array (MMA) and with the infrared-optimized 8-m-diameter telescope to be built on top of Mauna Kea, Hawaii, will measure features as small as a few times larger than an Astronomical Unit (or an AU, as the earth-sun separation is called by astronomers) at the distance of the nearest star-forming regions. All of these telescopes will help determine the temperature, density, and composition of protostellar disks. A technique called interferometry will be used at infrared wavelengths to discern angles as small as a thousandth of an arcsecond.[1] Images with this ultrahigh resolution may reveal traces of the formation of Jupiter-sized planets, such as gaps in the distribution of disk material.

[1] Angles are measured in degrees, which can be divided into 60 arcminutes and further subdivided into 3,600 arcseconds. The full moon subtends half a degree, and a large lunar crater an arcminute. An arcsecond is the angle subtended by a penny at a distance of two and a half miles. A thousandth of an arcsecond is the angle subtended by a penny seen from a continent away.

The Search for Other Planets

The Italian philosopher Giordano Bruno argued that space is filled with infinite numbers of planetary systems, inhabited by a multitude of creatures. For this and other indiscretions, Bruno was burned at the stake in 1600. Yet the question remains: are there other planetary systems in the universe? In the 1980s, for the first time, some evidence was found for disks of material surrounding other stars. However, as yet, no definitive observations have been made of a planet in orbit around another star. Some of the instruments proposed for the 1990s have the capability for detecting the Jupiter-sized planets of other stars.

Direct imaging of a distant planet is not easy, because the light from the parent star is billions of times brighter than the planet and because the star and its orbiting planet appear so close together as to be almost inseparable. The problem is analogous to trying to find from a distance of 100 miles a firefly glowing next to a brilliant searchlight. Fortunately, there are other ways to find planets. The position of the central star of a planetary system should wobble slightly in response to the changing gravitational tugs of its orbiting planets. At the distance of Alpha Centauri this shift in position amounts to a shift in angle of only a few thousandths of an arcsecond for a planet the size of Jupiter. Yet optical and infrared telescopes linked together as ground- or space-based interferometers will be capable of measuring such small angles and of surveying hundreds of stars within 500 light-years for the presence of distant planets like our own Jupiter. The orbital wobble of parent stars should also produce small velocity shifts that should be detectable with sensitive instruments on the large ground-based telescopes to be built in the 1990s.

Comets and the Origins of Life

Our sun and its solar system were formed about 4.5 billion years ago, as deduced from the proportions of uranium and lead measured in meteorites. In the intervening eons, life formed, and, most recently, humans began to explore the universe. What molecules were present to make the amino acids, proteins, and DNA that formed the first living creatures? What was the origin of life?

Weather, volcanoes, and bombardment by asteroids have erased from the planets almost all traces of the initial conditions in the solar system. Comets, however, spend most of their time far from the sun and may show us the pristine material of our world. Astronomers were surprised to find that the carbon in Halley's Comet (Figure 2.1) exists in the form of "tar balls" of complex organic molecules, rather than in simpler methane and carbon monoxide gas. The biological significance of this discovery is still hotly debated.

Comets will be closely studied in the coming decade. In addition to the close-range exploration planned for NASA's Comet Rendezvous Asteroid

FIGURE 2.1 A photograph of the nucleus of Halley's Comet taken by the Giotto spacecraft. Comets are thought to contain samples of the primordial material out of which the solar system formed 4.5 billion years ago. Complex organic molecules found in the material boiled off from Halley. Reprinted by permission from H.U. Keller. Copyright © 1986 Max-Planck-Institut für Aeronomie.

Flyby spacecraft, observatories on the ground such as the MMA and new 8-m-diameter optical and infrared telescopes, in airplanes such as the Stratospheric Observatory for Far-Infrared Astronomy (SOFIA), and in orbit about the earth such as SIRTF will determine the size, distribution, and composition of comets as far away as the orbit of Jupiter.

Weather and Volcanoes

What forces create the climate and weather on the earth? What forces make the continents drift and volcanoes erupt? Is our climate changing because of the actions of humans? Some information on these questions comes from measurements of the variability of the sun's output and of how other stars like the sun vary. Different perspectives come from observing other planets and their satellites with robotic spacecraft sent on voyages of discovery, and from telescopes on, or in orbit around, the earth. Observations of the atmospheres of the other planets made possible by infrared and radio telescopes have revealed the constituents of those atmospheres, as well as variations in their temperature and density with height. Careful examination of the dimming of the light from a star as it passed behind the planet Pluto was used to probe the atmosphere of this distant planet. Future observations will probe the climatology and meteorology of the planets and their satellites to give a comparative basis for understanding our own environment.

The Voyager spacecraft found sulfur-spewing volcanoes on Jupiter's satellite Io that appear to be driven by the grinding tides raised by the giant planet. These volcanoes are now monitored by earth-bound telescopes using cameras sensitive to the heat, or infrared radiation, from the volcanoes (Plate 2.2). The sizes, shapes, and compositions of asteroids are revealed by infrared observations and by bouncing strong radar signals off these objects (Plate 2.3).

THE LIFE HISTORY OF STARS

The Sun

In his masterwork the *Principia* (1687), Newton wrote that "those who consider the sun one of the fixed stars" may estimate the distance from the earth to a star by comparing its apparent brightness with that of the sun—in the manner that the distance to a candle may be judged by comparing its brightness with that of an identical candle nearby. Newton then calculated that the closest stars are about a million times farther away than the sun, in good agreement with later measurements. The sun is the closest star, and its careful study has important implications for our knowledge of all stars.

The sun is a great ball of hot gas, about a million miles in diameter. According to modern theory, its central density is about 100 times that of water, and the temperature is about 15 million degrees celsius. Such high temperatures are needed to smash subatomic particles together violently enough to fuse them and release nuclear energy. The liberated energy does two things. It maintains the heat within the sun, providing sufficient pressure to resist the inward pull of gravity. The liberated energy also turns into radiation, which slowly makes its way to the solar surface, finally creating the light we see. Some of the sun's energy goes into churning up its surface and producing extremely energetic particles, magnetic fields, solar flares, and a tenuous atmosphere of higher temperature called the corona.

Many mysteries remain. How do the magnetic fields get their energy? What is the role of magnetic fields in the curious 11-year cycle of activity

of the sun? Astronomers from around the world are proposing to build the Large Earth-based Solar Telescope (LEST), a 2.4-m solar telescope in the Canary Islands. The hallmark of LEST is unprecedented angular resolution, obtained by the use of special optics that will remove the distorting effect of the atmosphere. The principal aim of LEST is to observe the solar surface with sufficient acuity to reconstruct the three-dimensional structure of solar magnetic fields.

In conjunction with LEST, orbiting satellites equipped with visible, ultraviolet, and x-ray telescopes will study the radiation from the sun. Other x-ray satellites, such as the Advanced X-ray Astrophysics Facility (AXAF) already under construction, will monitor the coronal emission from other stars to help understand our own sun.

Astronomers have little direct data from the interior of the sun or any other star. The light we see comes from its surface. However, the surface motions contain clues about conditions far below. A major new field of study called helioseismology measures these vibrations of the solar interior. Just as the intensity and intervals of terrestrial earthquakes tell us about conditions deep within the earth, so also do the vibrations of the sun's surface inform us about the density, temperature, and rate of rotation in the deep interior (Plate 2.4).

A better knowledge of the interior of the sun might resolve another problem that has worried astronomers for years. The number of subatomic particles called neutrinos emitted by the sun and detected on the earth is much smaller than predicted. Neutrinos are produced in nuclear reactions at the center of the sun, and their rate of production has been calculated from our theories of nuclear physics and presumed knowledge of the conditions of temperature and density in the sun. Since neutrinos interact extremely weakly with other matter, almost every neutrino produced at the sun's center should escape. For the last 20 years, American physicists have counted neutrinos emitted from the sun and found fewer than one-third of the predicted number. The discrepancy is serious. Either the interior conditions of the sun are not what we think, or the neutrino has some property that allows it to change form and avoid detection once emitted. The former explanation, if correct, could alter our theories of the structure of stars. The latter would have significant implications for our understanding of subatomic physics. Planned for the 1990s are several new, more sensitive experiments to monitor neutrinos emitted from the sun.

The Formation of Stars

It takes two things to form a star—matter, and a mechanism to compress the matter to high density. Matter is plentiful in space in the form of diffuse hydrogen gas along with traces of other elements and small particles of dust. A dense clump within a gas cloud can pull itself together by gravity, becoming even denser. When the inward pull of gravity is sufficiently strong to overcome the pressure that tends to blow the clump apart, then the cloud will contract and fall toward its center. In a rotating cloud, a disk of gas and dust the size of a solar system may form in orbit around the nascent star. Matter continues to accrete onto the still-contracting central object, now called a protostar. As the cloud and protostar contract, gravitational energy is released as heat, and the protostar glows brightly at infrared wavelengths. When the temperature at the center of the protostar rises to around 10 million degrees celsius, enough to ignite nuclear reactions, a star is born. Depending on the circumstances of their birth, stars have masses ranging from about 0.1 to 100 times the mass of our sun. Smaller masses never get hot enough at their centers to ignite nuclear reactions; larger masses blow themselves apart at formation by the outward force of their own radiation.

This theory of the formation of stars was given support in the 1980s, when the IRAS detected tens of thousands of stars in the process of formation. More specifically, the satellite detected embryonic stars enshrouded in dense cores of gas clouds, during the early phase of collapse before the nuclear reactions had begun (Plate 2.5). Our understanding of star formation was given another boost in the 1980s when telescopes operating at millimeter wavelengths made an unexpected discovery: streams of gas flowing outward in opposite directions as jets from the vicinity of embryonic stars. Theorists argue that these gaseous streams may be aligned by a planet-forming disk around the young star. These jets are observed at millimeter wavelengths and occasionally break out of the surrounding cloud to become visible at optical wavelengths (Plate 2.6).

Because the earliest stages of star formation occur within very dense clouds of gas, impenetrable to visible light, clues must be sought in the radio waves and infrared radiation that can escape the thick clouds (Plate 2.7). During the 1990s, star formation will be a major focus of study with many of the telescopes proposed in Chapter 1, including SIRTF, SOFIA, the MMA, and the infrared-optimized 8-m telescope on Mauna Kea. Each of these telescopes will make unique contributions to these studies. SIRTF will make sensitive spectroscopic measurements at wavelengths inaccessible from the ground. SOFIA will observe submillimeter wavelength radiation from a large variety of molecules and atoms to characterize the conditions of high density and temperature in protoplanetary disks. The MMA and the infrared-optimized 8-m telescope will make high-spatial-resolution studies of the disks and the outflowing jets that will clarify the dynamics of these regions (Plate 2.8).

The systematic, thorough investigation of star formation throughout the galaxy is important, because star formation and the power of hot, young stars is an important energy-generation mechanism in many galaxies. IRAS found that some galaxies emit copious amounts of infrared energy, perhaps due to bursts of star formation thousands of times more intense than anything seen in our own galaxy. We can study the basic physical processes of star formation up close in our own galaxy and apply that knowledge to more distant systems.

The Life and Death of Stars

Once a star has formed, it spends most of its active lifetime burning its initial nuclear fuel, hydrogen, by a process of nuclear fusion that combines four hydrogen atoms together to make a helium atom and energy. Eventually, the star's hydrogen supply is spent, and, if the star is massive enough, the star's helium supply begins to fuse together to make carbon atoms, which then go on to make heavier and heavier atoms until all the nuclear fuel is converted to the element iron, which is incapable of further energy-releasing reactions. Once there exist no further resources of heat and pressure to counterbalance the inward pull of gravity, the star must collapse.

Our own sun has already lived about 5 billion years and will live another 5 billion years, quietly burning its hydrogen, before it swells into a red giant star. Then, in the relatively brief period of about 100 million years, it will exhaust the rest of its nuclear fuel and collapse. More massive stars spend themselves more quickly and less massive stars more slowly. For example, a star of 10 times the mass of our sun burns up its core of hydrogen gas and becomes a red giant in only about 30 million years.

A burned-out star may end its life in several ways. If the mass remaining after the red giant phase does not exceed several times that of our sun, it becomes a dense, dim white dwarf star or, after a violent stellar explosion called a supernova, an even denser cold star called a neutron star. A white dwarf is about 100 times smaller than a younger star of the same mass. A neutron star is yet another 1,000 times smaller than a white dwarf and is composed almost entirely of neutrons, uncharged subatomic particles, packed together side by side. A typical neutron star has an incredible density: the mass of Manhattan Island squeezed into a cherry. Furthermore, that star can spin very rapidly, between 1 and 1,000 revolutions per second; it can anchor magnetic fields that are trillions of times stronger than the earth's, and it can produce periodic pulses of intense radio waves. White dwarfs and neutron stars support themselves against further collapse by the resistance of their subatomic particles to being squeezed more closely together. They can remain in such balance almost forever. Yet these massive spheres that were once shining stars have no source of energy, other than their energy of rotation, and so eventually grow dim and cold.

The first white dwarf was identified in 1914 and the first neutron star in 1967. Astonishing as it may seem, astrophysicists had predicted the characteristics of neutron stars before their discovery. Swiss-born astronomer Fritz Zwicky and German-born astronomer Walter Baade, working together in California, correctly forecasted the existence and properties of neutron stars as early as 1933, only two years after the discovery of the neutron itself in terrestrial laboratories. Such accurate predictions testify not only to the power of theoretical calculations but also to the validity of the assumption that the same physical laws found on the earth apply to distant parts of the universe.

White dwarfs are the corpses of stars whose initial mass was less than about eight times the mass of our sun. More massive stars that have exhausted their nuclear fuel face a different end. For such stars, no amount of resistive pressure can stave off the overwhelming crush of gravity. These stars will explode in a brilliant supernova, which for a brief time can shine with the power of 100 billion stars. The core remaining after the explosion will become a neutron star, or, if the core is massive enough, an object called a black hole, whose gravity is so intense that not even a ray of light can escape it. The intense bursts of radiation given off by magnetized, rapidly rotating neutron stars are often detected by radio telescopes as pulsars. Stellar black holes are inferred to exist from the x-ray emission seen from some binary stars.

In early 1987, astronomers were handed a rare opportunity to test their theories of the evolution, collapse, and explosion of stars. A star exploded nearby, without warning, offering an unprecedented view of a supernova. By carefully monitoring the light from Supernova 1987A, as it is called, and by identifying in older photographs the star that blew up, astronomers have learned a great deal about the origin and nature of supernovae. The event confirmed the general theoretical outlines. The infrared and gamma rays from the radioactive decay of cobalt, and the amount of nickel and other elements ejected by the explosion, could all be understood. Also detected from Supernova 1987A were neutrinos, whose properties are not completely known. According to previous theories, neutrinos should be manufactured in great numbers during the formation of a neutron star. The neutrinos detected on the earth from Supernova 1987A not only confirmed the predicted temperatures and densities inside a supernova, but they also allowed physicists to learn more about the neutrino.

Supernovae play a vital role in the life cycle of stars. The debris from stellar explosions spreads out into space and adds new ingredients to the gas between stars from which new stars form. Thus supernovae are beginnings as well as ends. Theoretical calculations suggest that essentially all of the chemical elements except hydrogen and helium, the two lightest elements, were manufactured by nuclear fusion inside stars. The vast majority of the 100 chemical elements, including oxygen and carbon and other elements that earthly life depends on, were synthesized in stars and blown into space. Some of this seeding occurs during the red giant phase, as a star sheds its surface layers, and some of it occurs in the wind of particles that flow from the hot stellar atmosphere. The rest happens in supernova explosions. Later generations of stars, such as our sun, are born from the gas enriched by these new elements. The gas between stars connects the generations, receiving from the old stars and giving to the new.

Supernova explosions are thought to be the underlying cause for the acceleration of protons and heavier atoms to extremely high energies. These particles, called cosmic rays, move at nearly the speed of light, gyrating in the magnetic field of the galaxy, to bring us news about distant parts of the galaxy, news that we can currently decipher only imprecisely. Some cosmic rays enter the earth's atmosphere and can be detected by telescopes that look for the light given off when cosmic rays collide with stationary molecules in the atmosphere. Yet an innovative telescope called the Fly's Eye (Plate 2.9), which consists of hundreds of photosensitive tubes that scan the skies over Utah, has found tracks of cosmic rays of such high energy that even supernova explosions may have been inadequate to accelerate them. If these cosmic rays come from outside the galaxy itself, nobody can explain why they are there. An enhanced Fly's Eye telescope planned for the 1990s may solve this question.

The causes of stellar explosions, and the enrichment by supernovae of interstellar gas with life-giving heavy elements, will be the focus of extensive study with many of the telescopes to be built in the 1990s. Measurements of cosmic rays can determine abundances of the elements in the galaxy directly. A satellite called the Advanced Composition Explorer will measure the abundances in cosmic rays of all the elements up to zirconium, an element 90 times heavier than hydrogen. Scrutiny of the x-rays and gamma rays emitted by supernova remnants will help identify the various kinds and proportions of atoms dispersed in supernovae (Plate 2.10). Such a task will be on the dockets of the Gamma Ray Observatory; an Explorer satellite equipped with a gamma-ray spectrometer; AXAF; and two ultraviolet instruments currently under construction, the Far Ultraviolet Spectroscopy Explorer (FUSE) and the Extreme Ultraviolet Explorer. Chemical elements manufactured and spewed out into space by supernovae can also be identified by their infrared emissions. The proposed SOFIA telescope, quickly deployable because it is aboard an aircraft, will have the flexibility to study supernova debris on short notice. The high sensitivity of the proposed SIRTF mission will permit infrared measurements of supernovae out to 30 million light-years from the earth.

THE LIFE HISTORY OF GALAXIES

The Milky Way as a Galaxy

The Milky Way, the faint opalescent band of light that sweeps across the night sky, is our galaxy of 100 billion stars, all orbiting in a flattened disk. The central portions of the Milky Way, obscured at visual wavelengths by intervening interstellar dust, are revealed by infrared observations from ground-based telescopes and by orbiting satellites. Clear views of the stars and gas clouds that form the disk of the Milky Way have been obtained by the IRAS and the Cosmic Background Explorer (COBE) satellite and are shown on the cover of this report and as Plate 2.11. The universe is filled with more distant galaxies, each containing billions of stars and gas in orbit about its center. It takes our sun about 250 million years to complete one orbit about the center of the Milky Way. Galaxies come in a variety of shapes. Some are nearly spherical, while others, like the Milky Way, are flattened disks with a bulge in the middle. Individual galaxies are separated from one another like atolls in a Pacific archipelago; clusters of galaxies are like individual island chains scattered across the vast ocean of nearly empty space.

The Evolution of Galaxies

It was only in the 1920s that astronomers realized that many of the fuzzy patches revealed by their telescopes were indeed distant assemblages of stars, that is, galaxies. For many years after their discovery, galaxies were assumed to be fixed and unchanging, but by the 1970s, astronomers realized that galaxies should, in fact must, evolve. Stars alter the chemical composition of interstellar gas in a one-way process that builds heavier and heavier atoms. Thus the chemical composition, color, and luminosity of a galaxy should all change in time. In addition, galaxies can evolve dynamically as giant galaxies cannibalize their smaller companions.

Finding direct evidence for the evolution of single galaxies is not easy. But, fortunately, light travels at a finite speed. Since the distances in space are large, we can use this effect to observe evolution directly. When we take a picture today of the Andromeda Galaxy, 2 million light-years away, we see that galaxy as it was 2 million years ago. When we look at a galaxy in the Virgo cluster of galaxies, 50 million light-years away, we see light that was emitted 50 million years ago. Looking deeper into space is looking further back in time. Telescopes are time machines. With larger telescopes, we can see more distant galaxies at earlier stages of evolution.

Unfortunately, the light from distant galaxies is faint. To detect such feeble light, astronomers need large telescopes equipped with sensitive detectors. With new telescopes and more sensitive electronic cameras, astronomers have begun to study galaxies at much greater distances and thus at much earlier stages of evolution. For example, some observational evidence suggests a systematic color change of galaxies with age, as theoretically predicted.

For the 1990s, astronomers are building several large, visible-light and infrared telescopes with diameters ranging from more than 300 in. (8 m) to nearly 400 in. (10 m), far larger than the 200-in. telescope at Palomar Mountain, California, which was completed in 1949. In the 1990s astronomers hope to see more distant galaxies, at a much earlier stage of evolution than any galaxies previously seen. What types of stars inhabit young galaxies? While individual stars are born and die, how does the bulk population of stars in a galaxy age in time? Does the shape of a galaxy change in time, or is it determined completely when a galaxy first forms? How does the total luminosity of a galaxy change in time? Like stars, galaxies are found in congregations. These congregations are called groups and clusters of galaxies. How do the neighboring galaxies in groups and clusters affect each other?

Astronomers believe that many galaxies went through an extremely energetic early phase of evolution in which almost all of their energy was produced in their centers. This belief is based on the discovery of quasars in the 1960s. On photographic plates quasars resemble stars, yet are as distant as and far brighter than entire galaxies. Evidently, an enormous amount of energy is produced in a tiny volume of space, perhaps as small as our solar system. Importantly, most quasars have been found far away. Since distance translates into time in astronomy, we can infer that most quasars lived and died in the distant past. Quasars are the dinosaurs of the cosmos. Astronomers theorize that quasars constituted the central regions of some galaxies at a very early stage of their evolution.

The new generation of large, visible-light and infrared telescopes and the already launched Hubble Space Telescope may be able to detect the weak light of infant galaxies harboring quasars and to advance the study of the connection between quasars and galaxies. New infrared telescopes, such as SIRTF and the ground-based, infrared-optimized 8-m telescope, will also play important roles in quasar research. Finally, new radio telescopes with extremely high angular resolution, particularly the Very Long Baseline Array, should be able to make radio-wave images of quasars themselves.

Great dust clouds apparently surround many quasars, absorbing their visible light and turning it into infrared radiation. In the 1980s, IRAS discovered extremely luminous galaxies emitting 90 percent or more of their energy as infrared radiation and apparently harboring quasars at their centers. Furthermore, many such galaxies appeared to be colliding with other galaxies (Plate 2.12). Could collisions of galaxies give birth to quasars, or refuel them? With its much greater sensitivity, SIRTF should be able to study the nature and evolution of these curious infrared galaxies. If a big fraction of quasars are produced by collisions of galaxies, SIRTF will make it possible for astronomers to find out.

The Power Source of Quasars and Active Galaxies

Quasars and active galaxies emit copious amounts of power across a broad range of wavelengths, from radio waves to x-rays. Where does their great energy come from? It certainly cannot come entirely from stars that radiate predominantly visible light. Furthermore, stars live on nuclear energy, converting matter into energy with an efficiency of less than 0.1 percent. Nuclear energy is not efficient enough to balance the huge energy budget of quasars and active galaxies. Finally, the stars in a galaxy are scattered about, while the energy of active galaxies is produced in a highly concentrated region smaller than a light-year in diameter. Even if a sufficient number of stars were somehow crammed into such a small volume, the resulting stellar system would be so dense that the stars would quickly collide with each other and coalesce into a single unstable, massive object.

For these reasons, most astronomers believe that quasars and active galaxies can be powered only by gravitational energy released at the center of the system. According to current ideas, a massive black hole, with a mass of a million to a billion times that of our sun, inhabits the middle of an active galaxy or quasar. Surrounding gas and stars fall under the gravitational grip of the central black hole. As gas plunges toward the black hole, it releases gravitational energy, which is then transferred into high-speed particles and radiation. Matter falling toward a black hole can convert 10 percent of its mass into energy before it enters the black hole and is never heard from again.

A key to understanding quasars and active galaxies is the mechanism for feeding gas to the central black hole. Is it constant or intermittent? What triggers it? Possible sources of gas include ambient gas in the central regions of the galaxy, the gravitational shredding of hapless stars that wander too close to the black hole, the disintegration of stars by collisions with each other, and the agitation of one galaxy by a close encounter or merger with another. For the more luminous active galaxies and the less luminous quasars, gas must be fed to the central black hole at the rate of about a sun's worth of mass per year. Isolated black holes, no matter how massive, produce very little energy. Thus an understanding of the environment of the central black hole and how it is fueled may be crucial to understanding why some galaxies are highly energetic and others are not.

How can we test the hypothesis of massive black holes? A massive black hole, even as massive as a billion times the mass of our sun, would have a diameter smaller than our solar system. At the distance of the nearest big galaxy, 2 million light-years from us, such a black hole would have an angular size of only a few billionths of a degree, too small to be seen by any telescope in the near future. However, a massive black hole might reveal itself by the way that it affects the motions and positions of surrounding stars. Trapped by the gravity of the hole, surrounding stars would huddle together closely and would hurtle through space more rapidly than if no black hole were present. Infrared observations by a telescope carried aloft in the Kuiper Airborne Observatory have revealed rapid orbital motions in the center of our galaxy, indicating the possible presence of a black hole with a mass of a few million solar masses. Hints of these effects have also been found in a number of nearby galaxies, and the Hubble Space Telescope will look at a larger sample of more distant galaxies.

Black holes might also be indirectly identified by the high-energy emission of the surrounding gas. It is believed that gas near a black hole orbits it in a flattened disk, similar to a protoplanetary disk, only hotter and much more massive. Some of the gas in the disk, heated to temperatures between 1 million and 1 billion degrees celsius, would radiate x-rays. Such x-ray emission will be studied by AXAF. Another characteristic feature of such high energies could be the production of electrons and their antiparticles, positrons. Once produced, particles and antiparticles annihilate each other in a burst of gamma rays. The Gamma Ray Observatory, to be launched by NASA in 1991, and other proposed gamma-ray detectors in space, will search for such gamma rays. Ground-based telescopes may have found gamma rays of extremely high energy coming from astrophysical objects: pulsars, x-ray binary stars, and black hole candidates. The gamma rays are detected by light or particles produced in showers when the gamma rays enter the earth's atmosphere. Experimental physicists are working on better ways to detect this radiation, and theorists are studying novel astrophysical mechanisms to explain its existence.

The mysterious "x-ray background" that has puzzled astronomers for 25 years may also yield its secrets to new x-ray satellites. They should be able to indicate definitely what part of radiation comes from hot gas, what part from distant quasars, and what part, if any, from still more exotic objects.

The dramatic and energetic behavior of active galaxies and quasars raises other questions. Some of the observed gaseous "jets" emanating at great speed from these objects are extremely narrow and well collimated. What produces and controls such columns of matter? Recently, some progress has been made on these questions. The jets radiate x-rays, visible light, and radio waves. Images made with large visible-light telescopes, radio telescopes, and the Einstein Observatory x-ray telescope of the 1980s have revealed exquisite details of the structures and blobs in the gaseous jets (Plates 2.13 and 2.14). In particular, radio observations with very high angular resolution show that new concentrations of gas enter into the jet every year or so. New radio interferometers should be able to investigate jets and the central sources at much higher resolution.

Continued theoretical work in the coming decade will also be crucial to understanding energetic jets. In the 1980s, some of the features of jets were reproduced in large computer simulations (Plate 2.15). In such simulations, the scientist programs the computer with the basic laws of physics describing how gas, radiation, and magnetic fields behave, sets up some initial configuration of matter and radiation, and then lets the computer calculate how the system evolves in time. Comparison to observation then guides refinements of the theory and suggests new observations.

The Birth of Galaxies

If we peer out far enough into space, we should see back to the epoch of galaxy formation. What should a young galaxy look like? Astronomers are not sure. Galaxies were probably formed about 10 billion to 20 billion years ago, perhaps a few hundred million years after the beginning of the universe itself. But little is known about infant galaxies and even less about galaxies being born.

Astronomers generally suppose that the mass in the universe long ago was smoothly spread about, but was bunched up very slightly here and there, like ripples on a pond. The origin of these initial ripples is still unknown. In a place where the mass was bunched up, gravitational forces were slightly stronger. This caused nearby mass to bunch up more, attracting more surrounding gas. The force of gravity then became even stronger, and the process continued until a stronger concentration of mass formed. For sufficiently large concentrated regions, the inward pull of gravity exceeded the outward force of pressure, and the region collapsed into a dense and coherent structure. Just as for individual stars, the collapse of a giant gaseous cloud to form a galaxy might be affected in later stages by the forces of gas pressure, radiation, and rotation as well as the force of gravity. In the last decade, a great deal of theoretical work has been done to understand competing models for the formation of galaxies.

Once galaxy formation is under way, do galaxies continue to grow by accumulating gas in their vicinity, or do they reach their final size rather quickly? What determines the shape of a galaxy? Why are some galaxies nearly spherical, while others appear as flattened disks? In spiral galaxies, like our own, does the central bulge form first and then the disk, or vice versa? What determines the odd features of some galaxies—the rings and warps and bars? Were these features built in at the beginning, or did they form later, as the result of gravitational forces within the galaxy? Or were they fashioned by a close encounter or merger with another galaxy?

In reaching their conclusions, theoretical astronomers have had little help from observations made to date, since no protogalaxy has yet been convincingly identified. The telescopes of the 1990s may change this state of affairs. A number of telescopes, including the ground-based 8- and 10-m telescopes, the orbiting Hubble Space Telescope with its infrared camera, and SIRTF, will all survey the sky for the faint wisps of light emitted by the first generation of stars forming in infant galaxies. Infrared wavelengths will be used, since the effect of the cosmic expansion is to shift the intrinsic blue light of young stars to redder and redder wavelengths.

Finally, important hints about galaxy formation lie in the gas between galaxies, called the intergalactic medium. As in the interstellar medium within individual galaxies, this gas is enriched with the various chemical elements manufactured within stars. And the intergalactic medium is the material out of which galaxies formed. A primary tool for analyzing the intergalactic medium has been the study of radiation from quasars. As this radiation travels from there to here, it passes through the intergalactic medium, and some of it is absorbed. The particular wavelengths absorbed indicate the chemical makeup of the intervening gas. How does the composition of the intergalactic gas change in time? Can this gas be used to date the epoch when galaxies first formed? Analysis of the intergalactic medium at large distances requires both a high sensitivity to dim light and the ability to disperse the incoming light into its component wavelengths. The needed abilities are beyond current ground-based 4-m telescopes but within reach of the Hubble Space Telescope and the 8- and 10-m telescopes of the coming decade.

Still further back in time the universe consisted of smoothly distributed, hot gas. It emitted radiation that we should be able to see today. In 1965, astronomers did discover a bath of radio waves filling all space. It is believed that this radiation has been traveling freely through space, cooling as it goes, since the universe was only about 300,000 years old. At that time, the enormously hot energy of the cosmic fireball dominated the mass of the universe. Imprinted on this cosmic background radiation should be a record of the distribution of cosmic matter at that time, well before the epoch of galaxy formation. Irregularities in the distribution of matter at that time should be detectable now. They would show up today as variations in the intensity of the radiation detected by our radio telescopes pointed in different directions—and indeed all theories of the formation of galaxies demand the existence of such variations.

So far, to our puzzlement, no variations have been observed. From the measurements of the COBE, an orbiting satellite launched in 1989, and from other experiments, astronomers have recently determined that any variations in the intensity of the cosmic background radiation must be less than several parts in 100,000. Some theories of galaxy formation have been demolished by this fact. Revised theories that require the existence of large amounts of so-far undetected matter predict variations 10 times smaller. In the next decade, detectors now being developed should have the sensitivity required to challenge the new theories by looking for temperature variations of 1 part in 1 million. If no variations are found at these increased sensitivities, then theoretical extragalactic astronomy will be thrown into crisis. Something will be seriously wrong—either with our theories of galaxy formation or with our understanding of the cosmic background radiation. From such confrontations of theory with observations, deeper understanding emerges.

THE LIFE HISTORY OF THE UNIVERSE

The Big Bang Model

As we look further and further into space, will we come to an edge of space or a beginning of time? If the universe had a beginning, how did it begin? Will it have an end? These are questions in cosmology, the branch of astronomy concerned with the structure and evolution of the universe as a whole.

Every culture has had a cosmology. Aristotle's universe had an edge of space, an outermost sphere upon which were fastened the stars. But the cosmos had no beginning, or end, of time. In *On the Heavens*, Aristotle wrote that the "primary body of all is eternal, suffering neither growth nor diminution, but is ageless, unalterable and impassive." The Judeo-Christian world view did away with eternity but maintained the idea of a cosmos without change. According to this tradition, the universe was created from nothing by God and has remained much the same ever since. Copernicus, who in 1543 demoted the earth to a mere planet in orbit about the sun, changed the way we look at many things, but not the Aristotelian belief in a universe both spatially finite and static in time. In 1576, Thomas Digges became the first Copernican to pry the stars off their crystalline spheres and spread them throughout an infinite space. Now, the universe was without limit in space. However, the universe was still viewed as unchanging in time. Newton argued the same view a century later. Individual planets moved in the sky, to be sure, but the universe as a whole remained the same from one eon to the next.

Modern theories of cosmology date back to Albert Einstein's 1917 theory of general relativity and the models of the Russian mathematician and meteorologist Alexander Friedmann, who found a solution of Einstein's theory of gravity that described a universe that began in a state of extremely high density and then expanded in time, thinning out as it did so. Friedmann's model eventually came to be called the "Big Bang" model.

In 1929, the American astronomer Edwin Hubble, after whom the Hubble Space Telescope is named, made what was perhaps the most important observational discovery of modern cosmology: the universe is expanding. The universe is not constant in time; it is evolving and changing. Specifically, Hubble concluded that the other galaxies are moving outward from us in all directions. Hubble discovered that the distance to each galaxy is proportional to its recessional speed. A galaxy twice as far from us as another galaxy is moving outward twice as fast. This quantitative result had been predicted for a homogeneous and uniformly expanding universe.

If the galaxies are moving away from each other, then they were closer together in the past. At earlier times, the universe was denser. Continuing this extrapolation backwards suggests that there must have been a definite moment in the past when all the matter of the universe was compressed together in a state of enormous density. From the rate of expansion, astronomers can estimate when this point in time occurred: about 10 billion to 20 billion years ago. It is called the origin of the universe, or the Big Bang. As mentioned earlier, radioactive dating of rocks and meteorites, begun a couple of decades before Hubble's discovery, suggests that the sun and the earth are between 4 billion and 5 billion years old. Thus with two totally different methods, one using the outward motions of galaxies and the other using the rocks underfoot, scientists have derived roughly similar ages for the universe. This success has been a powerful argument in favor of the Big Bang model. However, it is important

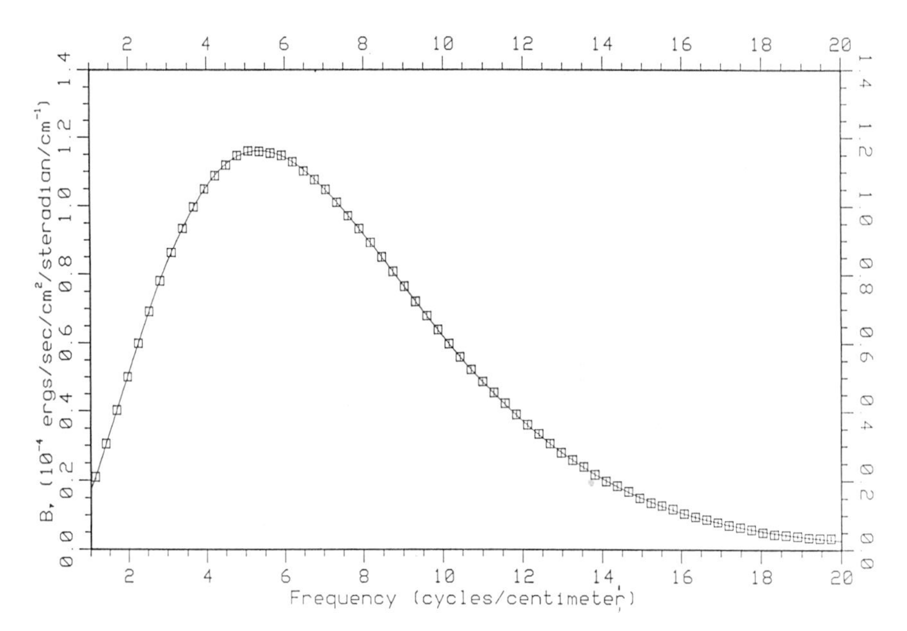

FIGURE 2.2 Spectrum of the cosmic microwave background measured by the far-infrared absolute spectrophotometer aboard COBE. This spectrum of the relict radiation from the "Big Bang" was taken by an orbiting satellite and depends on wavelength in the manner expected from a perfect emitter at a temperature of 2.74°C above absolute zero. Deviations from the theoretical profile might indicate the presence of galaxy formation at a very early age of the universe. Courtesy of the COBE Science Working Group.

to remember that cosmology, of all the branches of astronomy and indeed of all the sciences, requires the most extreme extrapolations in space and in time.

The Big Bang model, although widely accepted, rests on a rather small number of observational tests. In addition to explaining the observed expansion and age of the universe, the Big Bang model has successfully met two other major tests against observations. Calculations showed that the material of the universe should be, and measurements show that it is, approximately 74 percent hydrogen and 24 percent helium. The model also correctly predicts the abundances of other light elements such as deuterium and lithium.

In addition, the hot infant universe would have produced blackbody radiation that is easily identifiable by its universal spectrum (Figure 2.2). The most precise measurements of the cosmic background radiation have come from the COBE satellite, which has confirmed that the spectrum of the cosmic background radiation is extraordinarily close to that predicted by the Big Bang model.

According to the Big Bang model, hydrogen, helium, and some of the light elements, as well as the cosmic background radiation, were all created in the universe long ago, when the universe was very different from what it is

PLATE 2.1 An optical picture of the star Beta Pictoris taken with a special instrument that blocks out light from the central star. Light from the star is scattered by a disk of solid particles, perhaps a remnant of the formation of planets. Courtesy of B. Smith, R. Terrile, and the Jet Propulsion Laboratory.

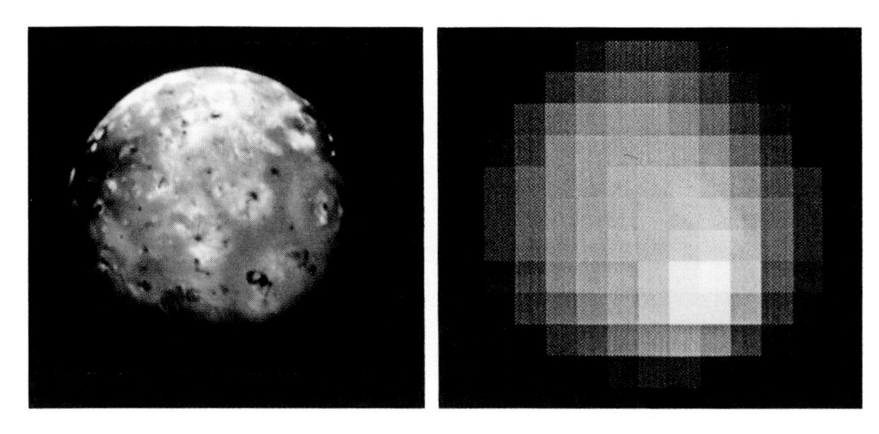

PLATE 2.2 Voyager spacecraft images of Jupiter's moon Io revealed the presence of sulfur-spewing volcanoes (left). Observations with an infrared camera on NASA's Infrared Telescope Facility (IRTF) on Mauna Kea made a diffraction-limited 3-μm image of the Loki, the brightest of these volcanoes (right). Courtesy of the Jet Propulsion Laboratory and the IRTF.

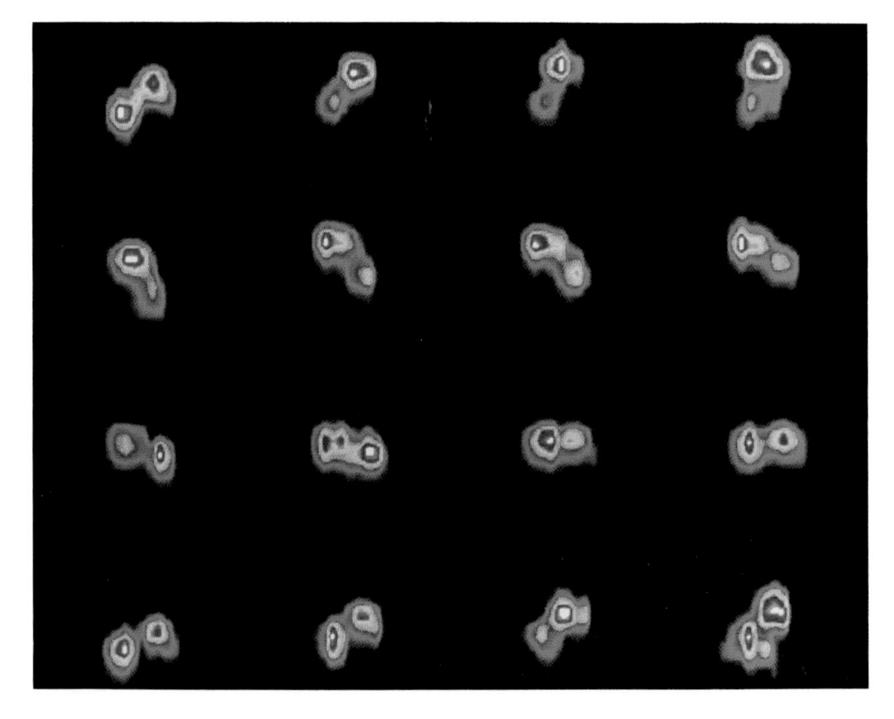

PLATE 2.3 Radar images of the asteroid 1989 PB made at the Arecibo Observatory near the time of closest approach of 2.5 million miles. The dumbbell-shaped asteroid is about a mile across and rotates with a period of about 4 hours. Courtesy of S. Ostro, Jet Propulsion Laboratory.

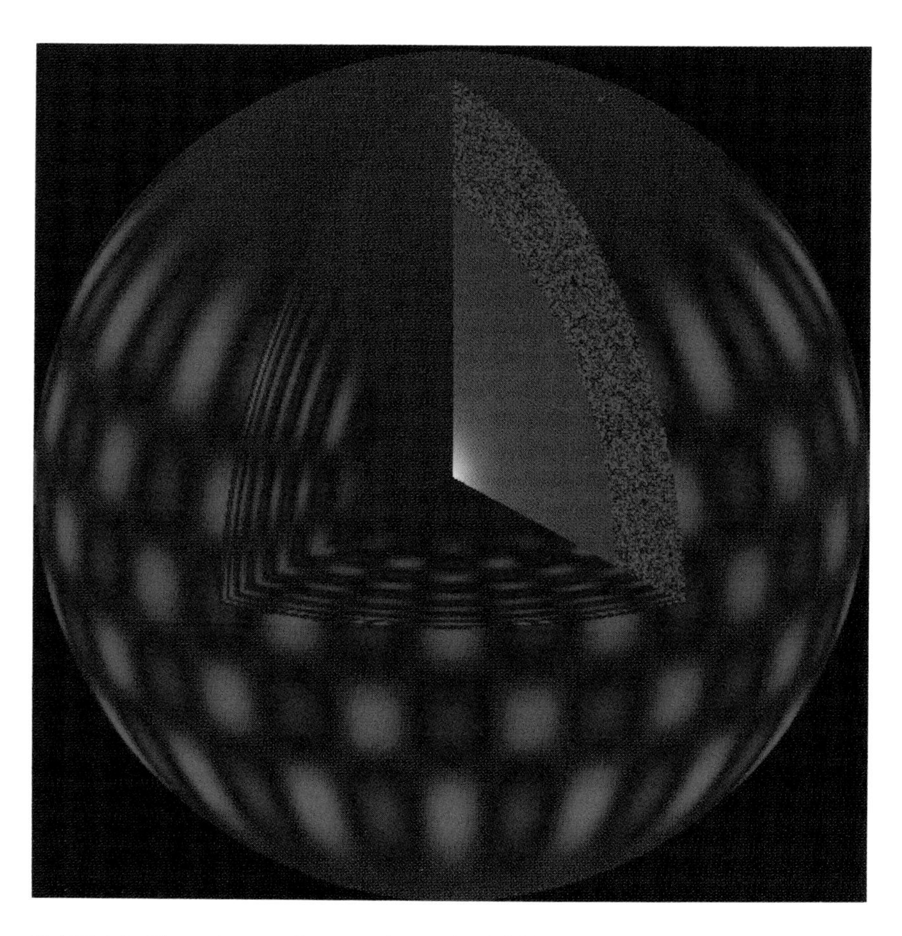

PLATE 2.4 The entire sun vibrates at thousands of different frequencies that penetrate deeply into the solar core. Astronomers study these vibrations in a manner analogous to terrestrial seismology to learn about the sun's interior. The figure shows an artist's representation of one of the sun's vibrational modes. Courtesy of the National Solar Observatory.

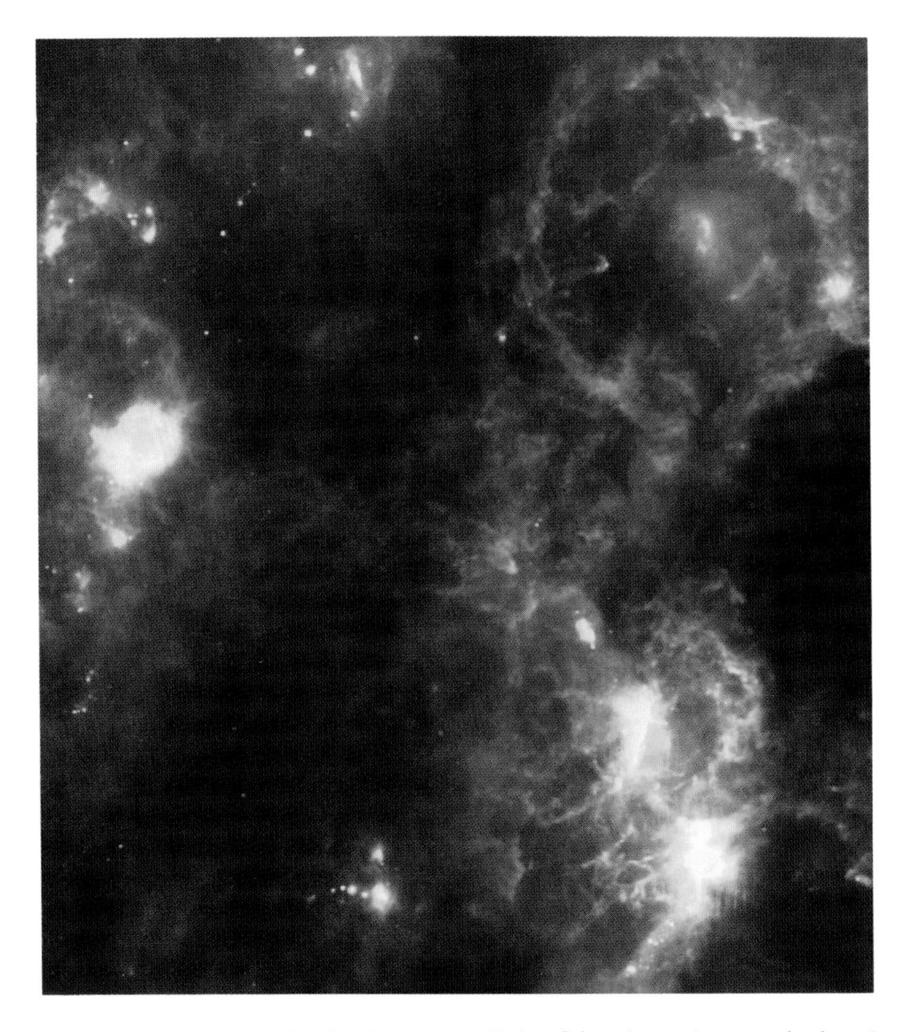

PLATE 2.5 A roughly 40° region in the constellation Orion shows giant gas clouds and hundreds of recently formed stars. The bright regions in the lower right are the Orion Nebula (Messier 42) and the Orion B complex where hot, young stars are being formed. This IRAS image shows the warmest material emitting at 12 μm as blue, cool material emitting at 60 μm as green, and the coolest material emitting only at 100 μm as red. Courtesy of the Infrared Processing and Analysis Center, California Institute of Technology.

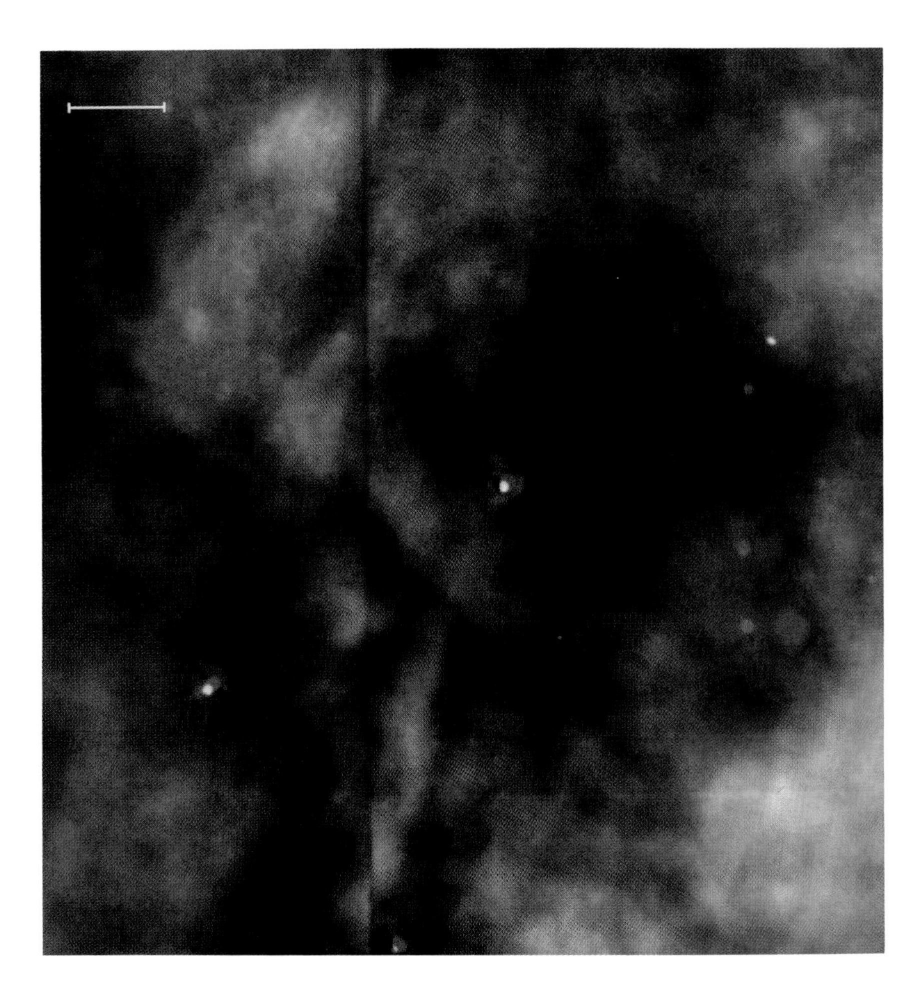

PLATE 2.6 A photograph made with the Wide-Field/Planetary camera on the Hubble Space Telescope shows a jet of material streaming away from a young star in the Orion Nebula, about 1,500 light-years away. Jets of this type are thought to be associated with the last stages of the accretion process by which stars are formed out of interstellar gas. The white bar at the top of the image represents 5 arcseconds. The smallest structures that can be resolved in this image are about one-tenth of an arcsecond, or roughly the size of our solar system. In this image red corresponds to radiation from ionized sulfur, green light to ionized hydrogen, and blue light to ionized oxygen. Courtesy of NASA.

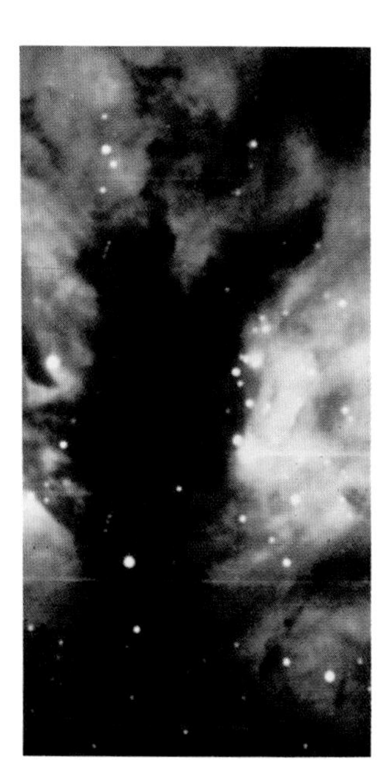

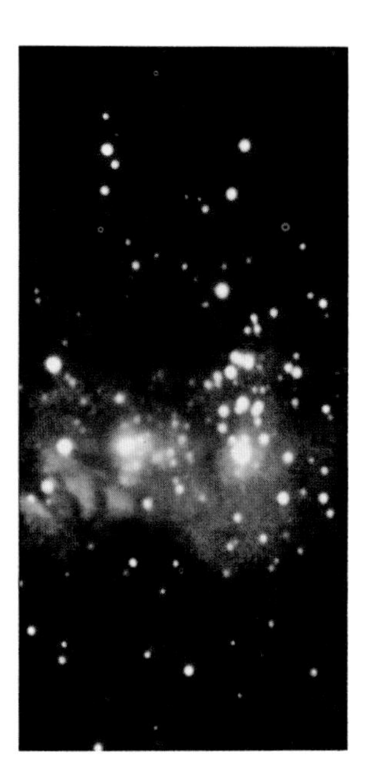

PLATE 2.7 Two views of the NGC 2024 star-forming region in the Orion Nebula. A picture of a 300-arcsecond field taken with a visible-light camera (left), showing relatively few stars due to obscuration by dust, and one taken with an infrared camera (right), showing hundreds of newly forming stars. Courtesy of the National Optical Astronomy Observatories.

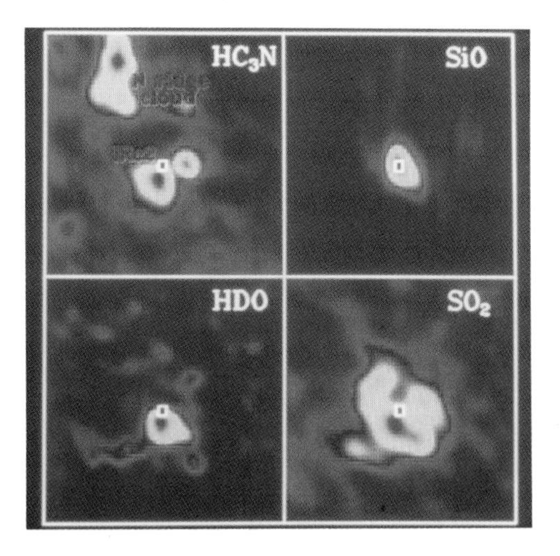

PLATE 2.8 Images taken with the Hat Creek millimeter interferometer show close-up views at a number of different frequencies of a star forming in the Orion Nebula. The images are 60 arcseconds on a side. Radio emission from the SiO, SO_2, HDO, and HC_3N molecules probes physical conditions in different regions close to the forming star, indicated by a square. Image courtesy of the Berkeley-Illinois-Maryland Array.

PLATE 2.9 The Fly's Eye telescope at Dugway, Utah, measures light produced in the atmosphere by very high energy cosmic rays. Each of the mirrors is equipped with 14 phototubes that allow the reconstruction of the trajectory of the cosmic ray. Photograph courtesy of Technical Information Department, Lawrence Berkeley Laboratory, University of California.

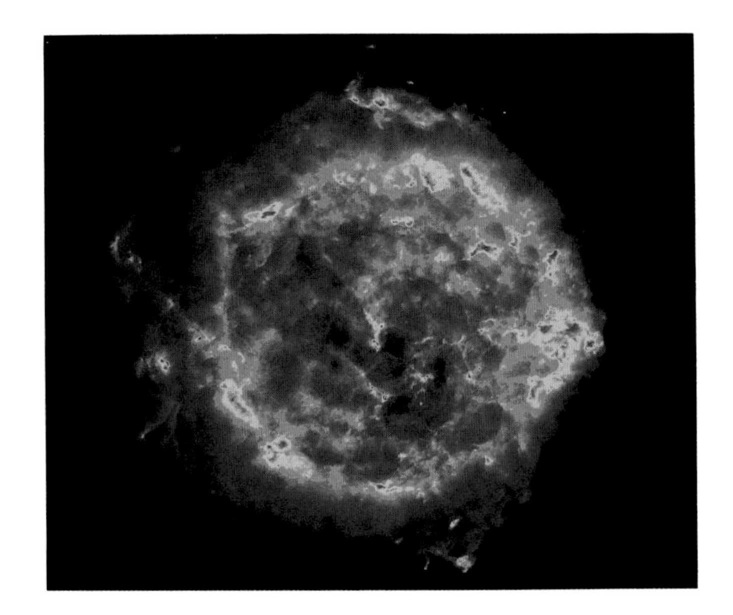

a

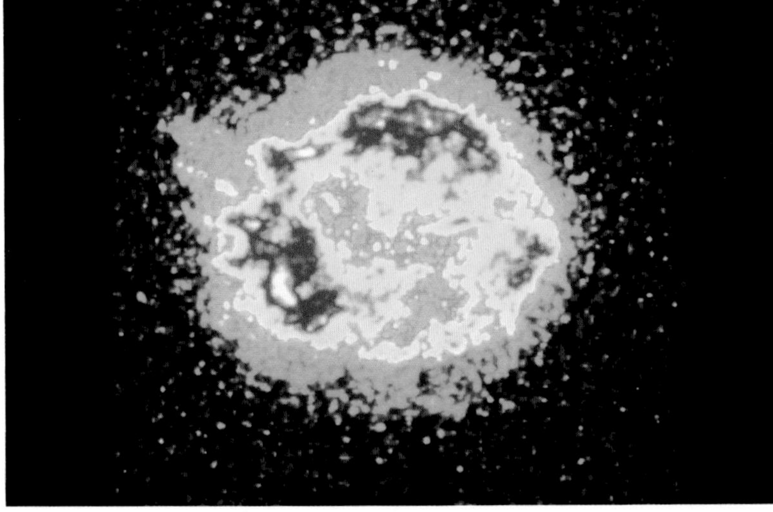

b

PLATE 2.10 Radio and x-ray images of the supernova remnant Cassiopeia A. (a) The radio image shows emission from electrons that have been accelerated by the interaction of the supernova blast waves and local magnetic fields. (b) The x-ray image reveals about 20 solar masses of debris from a stellar explosion at temperatures of 50 million degrees celsius, which has been expanding for about 300 years at speeds of ~5,000 km s^{-1}. Radio image courtesy of the National Radio Astronomy Observatory/Associated Universities, Inc. X-ray image courtesy of S. Murray.

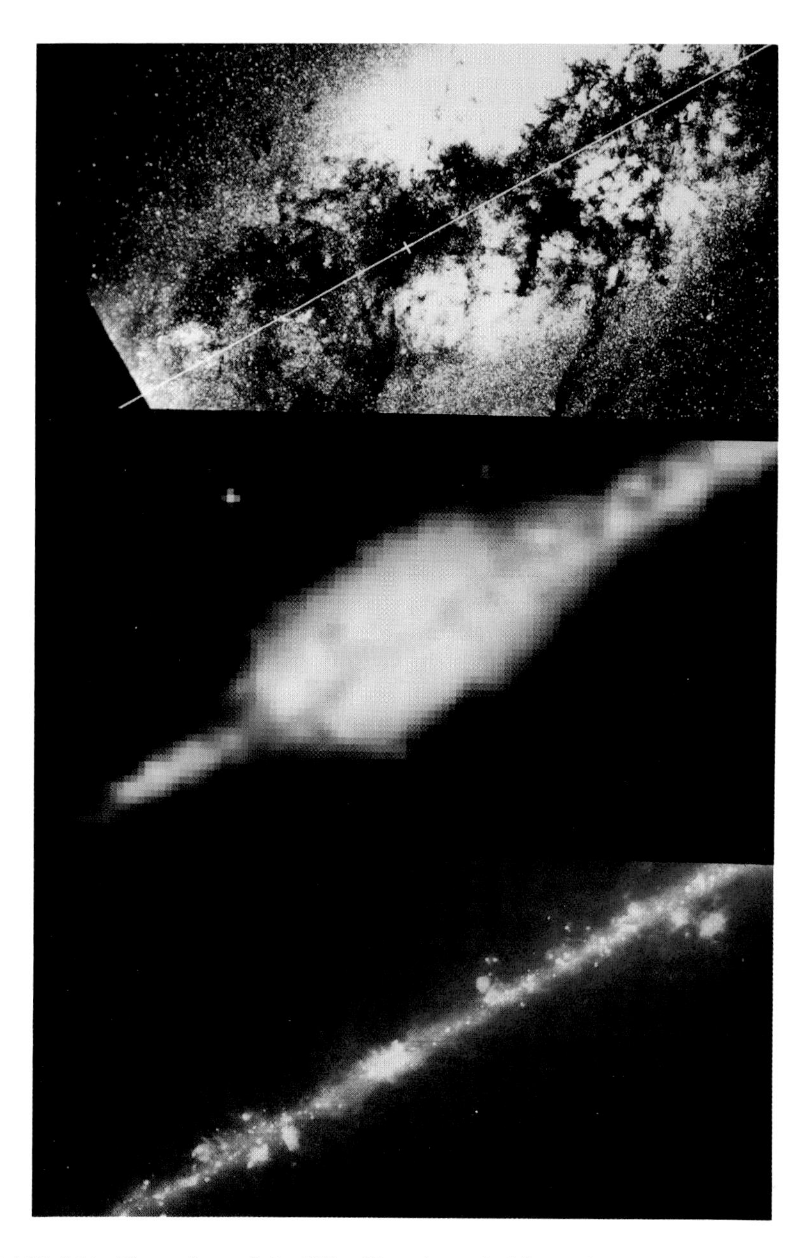

PLATE 2.11 Three views of the Milky Way. An optical image (top) shows mostly nearby stars. Clouds of gas and dust obscure the center of the galaxy, denoted by a (+). A near-infrared image (middle) from the COBE satellite penetrates the dust and shows the bulge of stars at the very center of the galaxy. A far-infrared image from IRAS (bottom) shows the cool, star-forming regions of the galaxy. Courtesy of the Palomar Observatory, Goddard Space Flight Center, and the Jet Propulsion Laboratory.

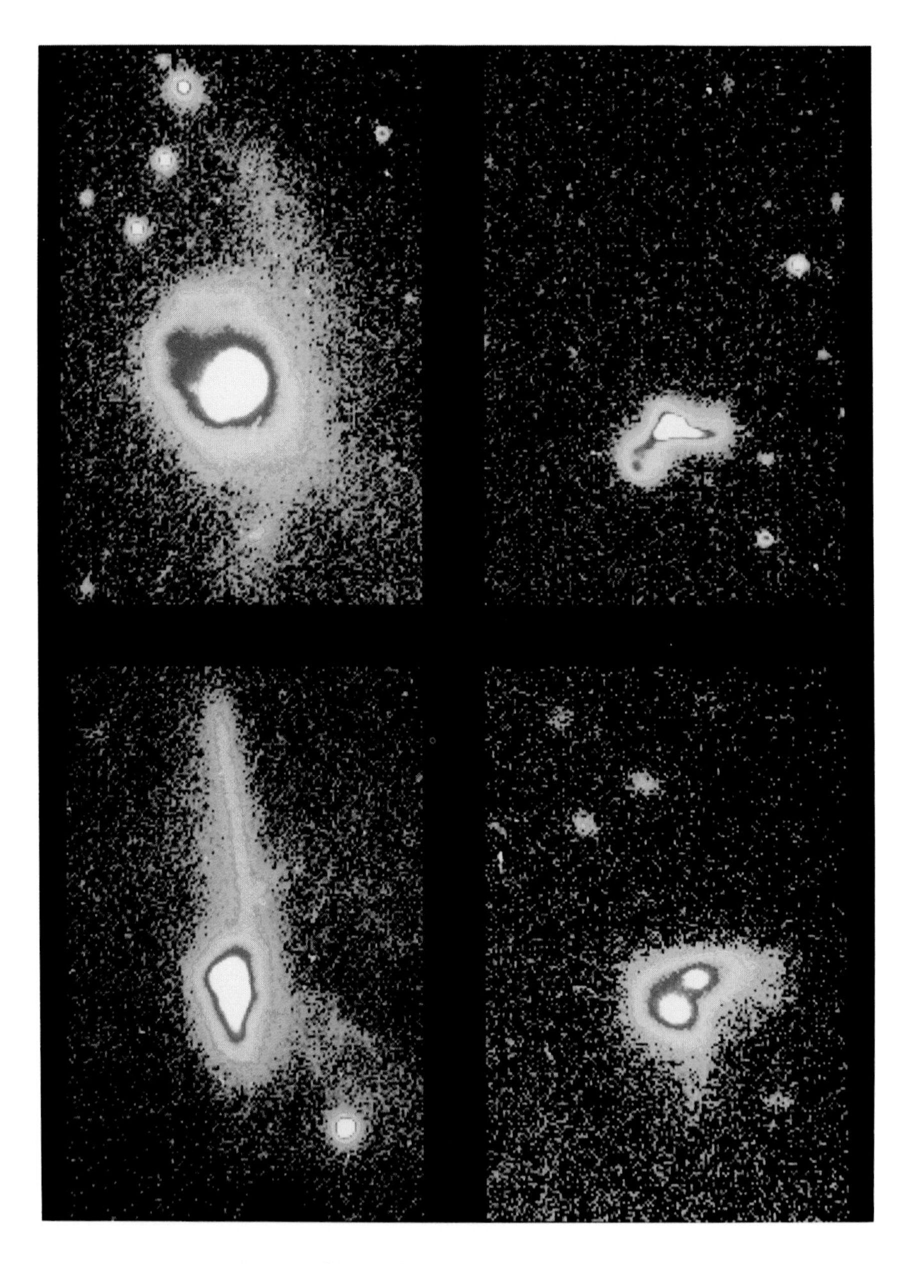

PLATE 2.12 Optical images of four infrared-luminous galaxies detected by IRAS show that many objects of this class are interacting or colliding galaxy pairs. Courtesy of D. Sanders and B.T. Soifer, California Institute of Technology.

PLATE 2.13 An image of the elliptical galaxy 3C353 shows visible light (blue) and radio emission (red). Jets of energetic particles moving at close to the speed of light emanate from the center of the galaxy and create two bright lobes of radio emission stretching over 360,000 light-years. Courtesy of the National Radio Astronomy Observatory/Associated Universities, Inc.

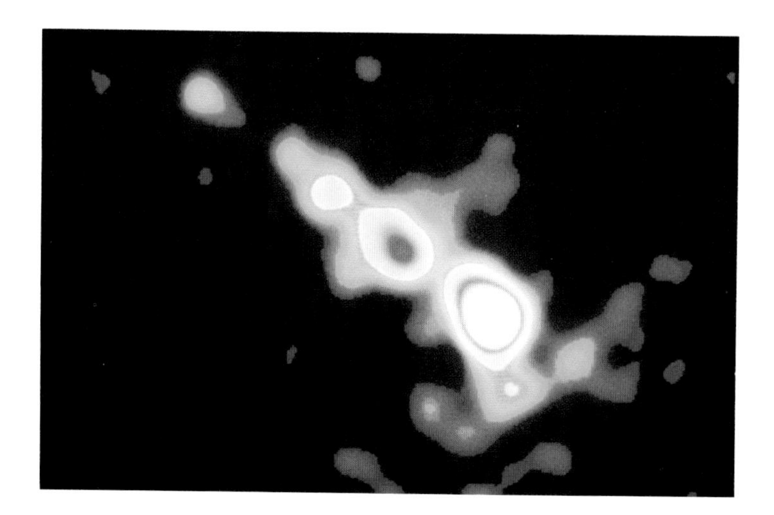

PLATE 2.14 X-rays from a "jet" in the active galaxy Centaurus A. X-rays are thought to come from a beam of very energetic particles emitted from the vicinity of a giant black hole in the nucleus of the galaxy. These particles, also detected at radio wavelengths, then stream out into intergalactic space. Image courtesy of E. Schreier.

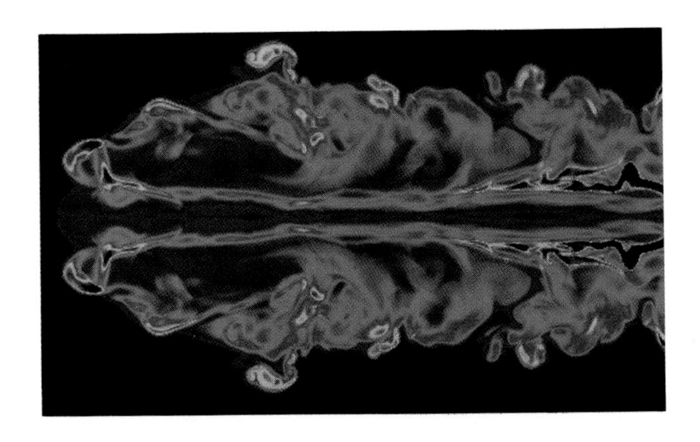

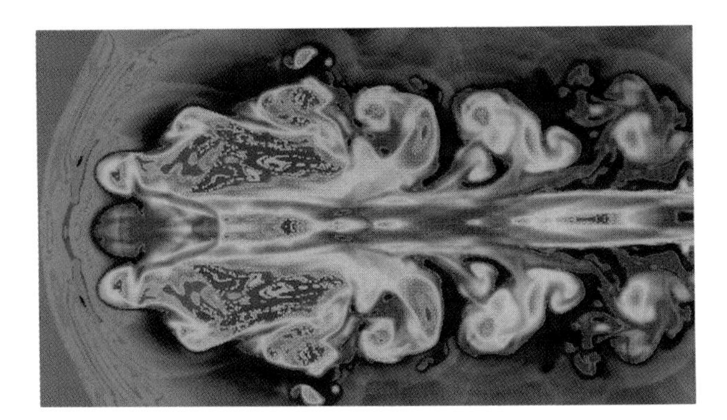

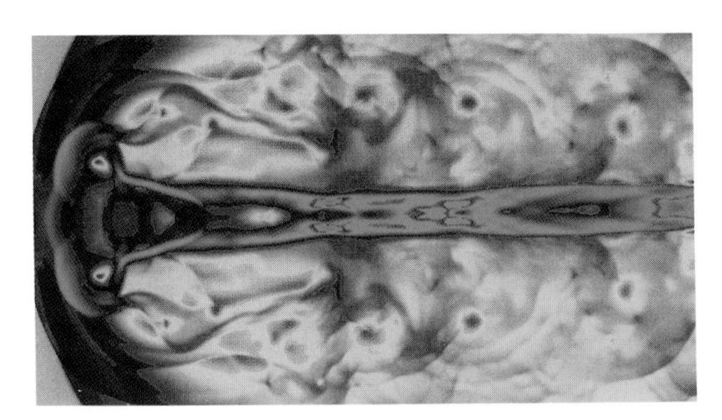

PLATE 2.15 A supercomputer simulation of a jet of material such as might be ejected from the center of a galaxy (see Plates 2.13 and 2.14). The top image shows the variation of total energy in the jet, the middle image the density of matter, and the bottom image the magnetic field strength. Courtesy of D. Payne and D. Meier of the Jet Propulsion Laboratory, K. Lind of the Naval Research Laboratory, R. Blandford of the California Institute of Technology, and T. Elvins of the San Diego Supercomputer Center.

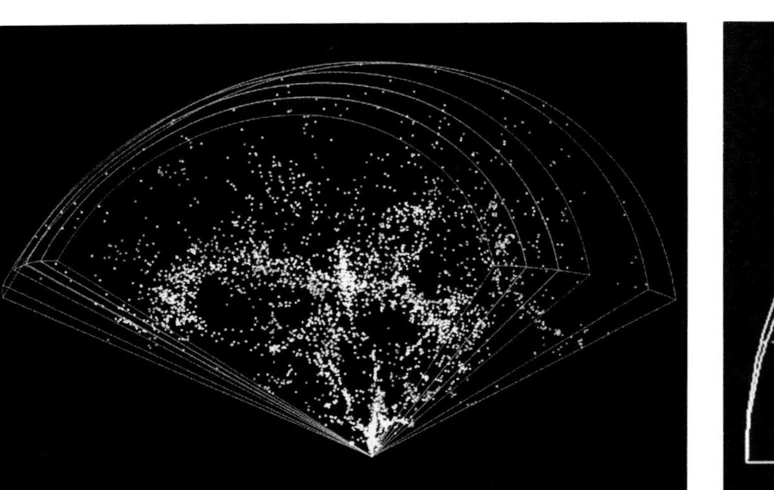

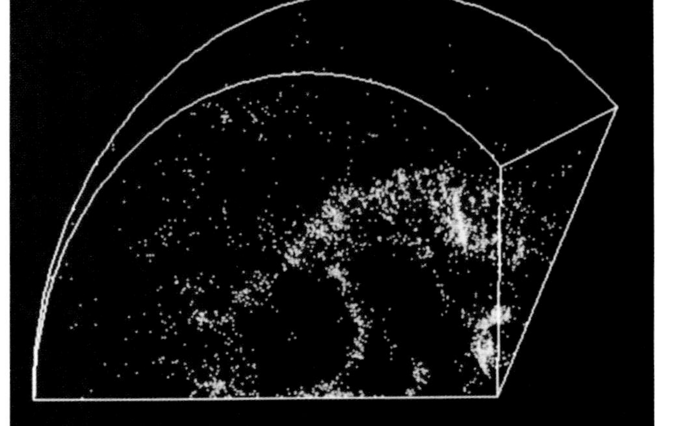

PLATE 2.16 The observed three-dimensional distribution of galaxies (left) shows sheets, filaments, and voids over 100 million light-years in size. Reprinted by permission from M. Geller and J. Huchra. Copyright © 1990 by the Smithsonian Astrophysical Observatory. Theoretical models (right) attempt to show how galaxies arrange themselves in such complex structures. Courtesy of C. Park and J.R. Gott.

PLATE 3.1 The Mirror Laboratory of the University of Arizona's Steward Observatory has developed innovative techniques for the casting and polishing of mirrors for astronomical telescopes. The 3.5-m mirror blank is a prototype for the 8-m mirrors to be built in the Mirror Laboratory in the 1990s. Courtesy of R. Angel and the Steward Observatory.

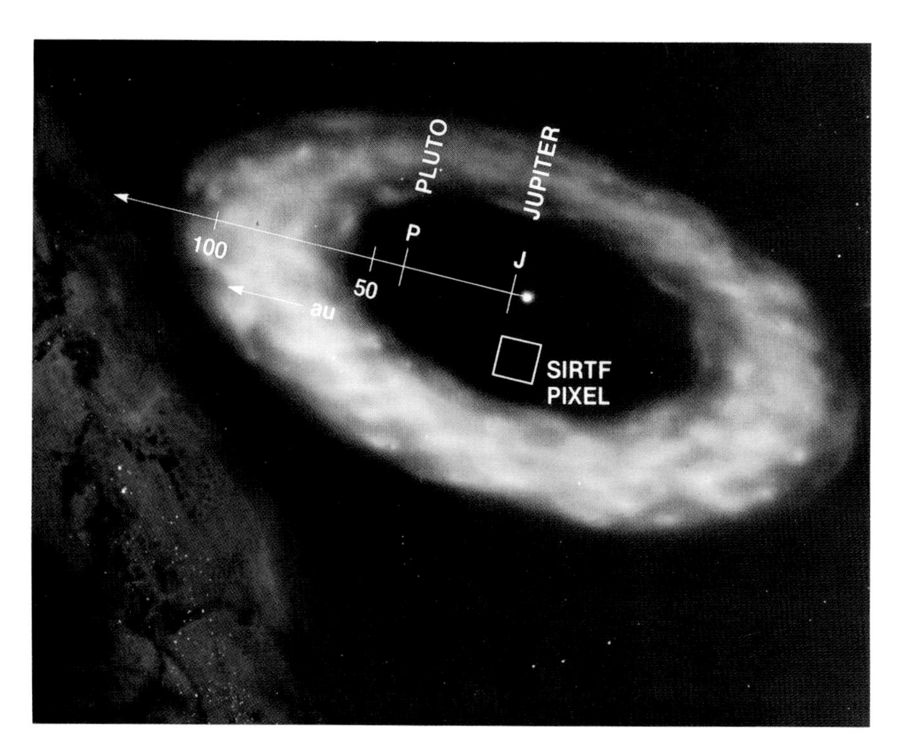

PLATE 4.1 An artist's conception of the solar-system-sized ring of planetary debris around a star like Vega or Beta Pictoris. SIRTF would be able to study the mass, composition, and distribution of matter in such systems out to 1,000 light-years away. The typical size of a pixel in an image from SIRTF is indicated.

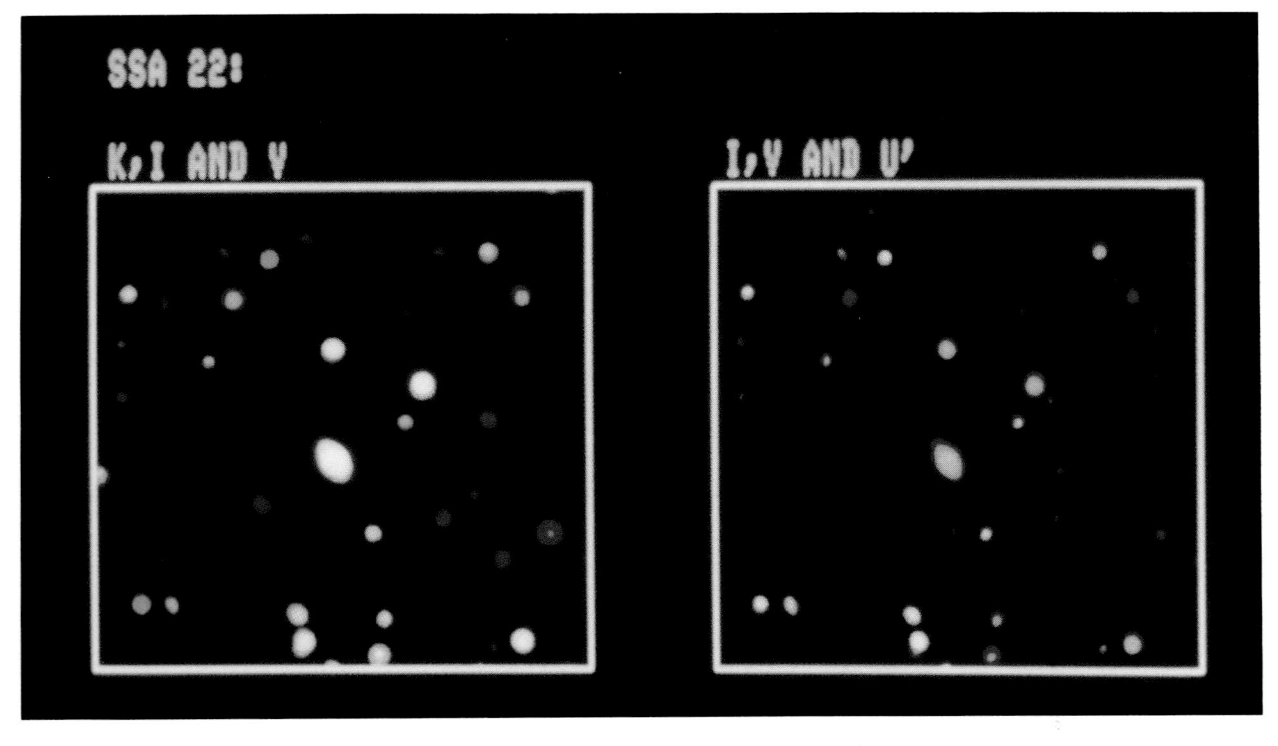

PLATE 4.2 A deep infrared survey looking for primeval galaxies. The image on the left was formed by superposing images taken in two visible wavelengths (0.55 and 0.9 μm in blue and green) and one infrared wavelength (2.2 μm in red). The image on the right was formed at three visible wavelengths (0.35, 0.55, and 0.9 μm). The images are 80 arcseconds on a side, and the infrared image required an exposure time of 22 hours to reach objects as faint as 21.4 magnitude at 2.2 μm. The reddest objects may be very distant spiral galaxies. Courtesy of L. Cowie, University of Hawaii.

today. Imagine a movie of cosmic evolution played backward in time, starting from the present. The universe contracts. The galaxies move more and more closely together. As the universe grows ever denser, stars and galaxies lose their identity, and the matter of the universe begins to resemble a gas. Like any gas being compressed, the cosmic gas becomes hotter and hotter. Eventually, the heat becomes so high that atoms cannot retain their electrons, and they disintegrate into atomic nuclei and freely roaming electrons. At a still earlier stage, closer in time to the Big Bang, the atomic nuclei themselves disintegrate into their constituents, protons and neutrons. At an even earlier time, each proton and neutron disintegrates into elementary particles called quarks. The universe becomes a roiling sea of subatomic particles. Very close to the Big Bang, the subatomic, or quantum, effects of gravity itself become important. At present, we have no theory that can handle quantum gravity, although some first steps have been taken in this direction. The concept of the Big Bang is the ultimate challenge to our understanding of physical laws.

The Large-Scale Structure of the Universe

The assumption of perfect homogeneity that underlies the Big Bang model is obviously not valid in the region we have observed with our telescopes. The universe nearby is not evenly filled with a smooth and featureless fluid. Rather, it is lumpy. Matter clumps into galaxies, and galaxies huddle together in groups and clusters of galaxies, and so on. Given the local lumpiness, the assumption of homogeneity means that space should appear smooth when averaged over a sufficiently large volume, just as a beach appears smooth when looked at from a distance of a few feet or more, even though it appears grainy when looked at from closer range.

Groupings of galaxies, which astronomers call "structures," are intriguing in their own right. Structures of various sizes abound, and astronomers want to understand the nature of these structures and how they were born. Were structures formed completely by gravitational attraction, or were other forces involved? What shapes do the structures form? How large is the largest structure? Until such questions are answered, it will be hard to decide whether the observed inhomogeneities and structures are simply details in the standard view or hints of a radically different picture.

Much evidence for inhomogeneities has been obtained in recent years, including the discoveries of chains of galaxies, sheets of galaxies, and giant voids with very few galaxies in them. In the late 1980s, American astronomers found evidence for the largest structure so far known, a "wall" of galaxies stretching at least 500 million light-years. The new three-dimensional maps of large samples of galaxies, produced from surveys of galaxy redshifts, have been made possible by advances in technology that allow the redshifts of galaxies to be measured with fast and automated procedures. Under the assumption that the universe is approximately homogeneous and uniformly expanding, the redshift of a galaxy translates into an approximate distance, thus providing the elusive third dimension.

It is not yet known whether the new cosmic structures found in a few selected regions of space are typical. What seems clear is that structures of some kind have been found at the largest possible scale in each survey of galaxies; that is, a survey that looks over a region of 100 million light-years usually finds some chain or wall or absence of galaxies extending roughly 100 million light-years in size; a survey of a 200-million-light-year region finds structures of 200 million light-years, and so on.

Galaxy surveys are now in progress that extend out to a few billion light-years. In addition, surveys with many more galaxies are being planned. The largest redshift surveys to date include only several thousand galaxies and sample only limited directions in the sky. In the coming decade, astronomers hope to initiate surveys of a million galaxies. Such surveys could be accomplished with moderate-sized visible-light and infrared telescopes. If in the future we find filaments and bubbles and voids with sizes of a few billion light-years, several times larger than those now mapped, then there would be a direct contradiction with the uniformity of matter implied by the cosmic background radiation. The Big Bang model might even be called into question.

On the theoretical side, astronomers are attempting to make sense of the observed positions and motions of galaxies by the use of large computer simulations. Such simulations involve 10,000 to several million particles, each representing a portion of a galaxy or a number of galaxies. The particles are placed at initial positions, given an initial outward velocity corresponding to the expansion of the universe, and then allowed to interact via their mutual gravity. The hypothetical galaxies fly around the computer screen, gravitate toward each other, and form clumps and wisps and voids. The largest simulations require the fastest and biggest computers in the world. By comparing computer simulations to the observed large-scale structure of the universe (Plate 2.16) scientists hope to test their assumptions about the initial conditions and forces at work in the cosmos. The current computer simulations, although 10 times larger than those of a decade ago, still do not have enough particles for a decisive comparison between theory and observation. Within the coming decade, larger computers and new methods for using those computers should give more reliable answers. And the new calculations will have more brain as well as brawn. Additional physics will be taught to the computers, and the resulting simulations will be more realistic and believable.

Whatever the outcome of the computer simulations, the ultimate theory of the distribution of matter in the universe must be consistent with all the observations. Astronomers have become increasingly worried about reconciling the smoothness of the cosmic background radiation with the lumpiness of matter nearby. Both theoretically and observationally, an understanding of the large-scale structure of the universe is at the top of most lists of the outstanding problems in cosmology.

Dark Matter

An obstacle to understanding the distribution of mass in the universe and the motions of galaxies is that at least 90 percent of the detected mass in the universe is not seen. No radiation of any kind—not visible light, nor radio waves, nor infrared, nor ultraviolet, nor x-rays—has yet been detected from this "dark matter." It is invisible to our instruments. We know that dark matter exists, because we have observed its gravitational effects on the stars and galaxies that we see, but we have little idea what it is.

The conundrum was uncovered in 1933 by Fritz Zwicky, who estimated the mass of a cluster of galaxies in orbit about one another by measuring the amount of gravity needed to hold the cluster together. He discovered that the total mass thus inferred was about 20 times what could be accounted for by the visible stars in the cluster. Zwicky's startling discovery was not appreciated by most astronomers until the 1970s, when it was confirmed by new observations of the orbital motions of stars and gas in individual galaxies. The amount of matter needed to produce these motions is far larger than the observed light-emitting matter. On the scale of both individual galaxies and of clusters of galaxies, most of the mass inferred to exist from the motions of stars and galaxies is not seen. Large ground-based optical and radio telescopes of the 1990s will be the instruments of choice to map the motions of stars and gas at the periphery of galaxies.

What is the nature of dark matter? Does it consist of numerous dark objects, like dim red stars, planets, or black holes, or does it consist of subatomic particles that interact with other matter only through gravity? Dark matter could alter our theories of the formation of galaxies or of subatomic particles. Whatever it is, astronomers have been startled by the realization that the luminous matter they have been staring at and pondering for centuries may make up a mere tenth of the inventory. And it is not just the unknown identity of dark matter that causes concern. Its quantity and arrangement in space are also uncertain, foiling attempts to understand why the luminous mass is arranged as it is.

A better understanding of how dark matter is distributed in space, particularly over the largest possible distances, may be the key to many of the puzzles discussed earlier. A careful reconciliation of the velocities of galaxies with the observed inhomogeneities in luminous matter should reveal the presence of dark matter, which contributes to the velocities through its gravitational effects. Detailed maps of the positions and motions of galaxies carried out on ground-based optical and radio telescopes may yield better maps of the location of dark matter.

Dark matter will also be mapped in the coming decade by x-ray emission from hot gas. Very hot gas has been detected inside large clusters of galaxies extending 5 million to 10 million light-years out from the center of many clusters. So hot that it should boil away, the gas is evidently held by the gravity of invisible matter. From the precise distribution of the gas, astronomers can work back to infer the gravity confining it and the distribution of dark matter producing that gravity. In the coming decade, the German X-ray Roentgen Satellite (ROSAT) already in orbit, the Japanese x-ray satellite ASTRO-D, and especially NASA's AXAF telescope will make better maps of the distribution and temperature of the hot gas in galaxy clusters.

The identification of dark matter may be a process of elimination. For example, the dark matter could be large planets, with masses between a thousandth and a tenth the mass of our sun. Such objects should have enough heat generated by their slow contraction to emit a low intensity of infrared radiation. A highly sensitive infrared telescope like SIRTF may be able to detect them. In particular, SIRTF will scrutinize the far reaches of our own galaxy, where dark matter may be lurking, and search for a faint excess glow.

There are other ways to probe dark matter. One of the most recent and potentially very important new techniques makes use of the "gravitational lens" phenomenon. When he published his new theory of gravity, Einstein pointed out that light, like matter, should be affected by gravity. Thus as light from a distant astronomical object, such as a quasar, travels toward the earth, that light should be deflected by any mass lying between here and there. The intervening mass can act as a lens, distorting and splitting the image of the quasar. Even if the intervening mass is totally invisible, its gravitational effects are not. By carefully analyzing the distortions of quasar images, astronomers can reconstruct many of the properties of the intervening gravitational lens, including its distribution in space and total mass. Gravitational lenses were first discovered only a decade ago; about a dozen have been found since that time. In the coming decade, the gravitational-lens phenomenon will be used as a powerful tool to uncover the nature of dark matter.

Alternatively, dark matter could consist of individual, freely roaming subatomic particles, rather than aggregates of particles such as planets. The possibilities have stirred the imaginations of particle theorists. Dozens of particles have been proposed, some on the basis of new theories of subatomic physics. None, however, have as yet been seen in the laboratory. If dark matter does indeed consist of these exotic particles, then it may be identified in the laboratory rather than at the telescope. Within the last few years, the first laboratory detectors have been built to search for some of these hypothesized particles. The experiments are extremely difficult, owing to the elusiveness of the particles, and it is estimated that detectors of the next decade need to be approximately 100 times more sensitive before the particles can be found—if they exist.

The Origin of the Universe

The goal of physics is to explain nature with as simple a theory as possible, and the goal of cosmology is to explain the large-scale structure and evolution of the universe in terms of that theory. Astronomers and physicists today believe that many properties of the present universe probably depend on what happened during the first instants after the Big Bang. One such property, ironically, is the apparent uniformity of the universe on the large scale, as evidenced by the cosmic background radiation. Although such uniformity and homogeneity have been assumed in the Big Bang model, they still must be explained, or at least be made plausible. It seems unlikely to many scientists that the universe would have been created so homogeneous by accident.

In the 1970s an important change occurred in theoretical cosmology. Physicists with expertise in the theory of subatomic particles joined astronomers to work on cosmology. The physicists brought a fresh stock of ideas and a new set of intellectual tools to bear on the question of why the universe has the properties it does, not just what those properties are. Particle physicists in the United States and in the Soviet Union proposed a modification to the Big Bang model called "the inflationary universe" that has caused a major change in cosmological thinking. The essential feature of the inflationary universe model is that, shortly after the Big Bang, the infant universe went through a brief and extremely rapid expansion, after which it returned to the more leisurely rate of expansion of the standard Big Bang model. By the time the universe was a tiny fraction, perhaps 10^{-32}, of a second old, the period of rapid expansion, or inflation, was over. The epoch of rapid expansion could have taken a patch of space so tiny that it had already homogenized and quickly stretched it to a size larger than today's entire observable universe. Thus the inflationary expansion would make the universe appear homogeneous over an extremely vast region, far larger than any region from which we have data.

The inflationary model makes specific predictions about the formation of structures in the universe. In particular, processes in the early universe would have determined the nature of the initial inhomogeneities that later condensed into galaxies and clusters of galaxies. Some of the predictions of this model will be tested by the COBE satellite and by galaxy surveys to be made in the 1990s with new optical telescopes. Whether these and other observations will confirm the inflation model or lead theorists to a different approach altogether is, of course, unknown. But astronomers and physicists will be working together to learn the past history of the observable universe.

The End of the Universe

As the universe expands, its parts pull on one another owing to gravitational attraction, and this slows down the expansion. The competition between the outward motion of expansion and the inward pull of gravity leads to three possibilities for the ultimate fate of the universe. The universe may expand forever, with its outward motion always overwhelming the inward pull of gravity, in the way that a rock thrown upward with sufficient speed will escape the gravity of the earth and keep traveling forever. Such a universe is called an open universe. A second possibility is that the inward force of gravity is sufficiently strong to halt and reverse the expansion, just as a rock thrown upward with insufficient speed will reach a maximum height and then fall back to the earth. A universe of this type, called a closed universe, reaches a maximum size and then starts collapsing, toward a reverse big bang. This universe grows smaller and hotter, and has both a beginning and an end in time. The final possibility, called a flat universe, is analogous to a rock thrown upward with precisely the minimum speed needed to escape from the pull of the earth. A flat universe, like an open universe, keeps expanding forever. Meanwhile, stars and galaxies evolve, heavy elements are synthesized, radioactive decay transmutes elements, and the universe grows colder and colder.

The Big Bang model allows all three possibilities. Which one holds for our universe depends on how the cosmic expansion began, in the same way that the path of the rock depends on the rock's initial speed relative to the strength of the earth's gravity. For the rock, the critical initial speed is 7 miles per second. Rocks thrown upward with less than this speed will fall back to the earth; rocks with greater initial speed will never return. Likewise, the fate of the universe was, according to the Big Bang model, determined by its initial rate of expansion relative to its gravity. Even without knowledge of these initial conditions, however, we can infer the fate of our universe by comparing its current rate of expansion to its current average density. If the density is greater than a critical value, which is determined by the current rate of expansion, then gravity dominates; the universe is closed, fated to collapse at some time in the future. If the density is less than the critical value, the universe is open. If it is precisely equal to the critical value, the universe is flat. The ratio of the actual density to the critical density is called omega. Thus the universe is open, flat, or closed depending on whether omega is less than 1, equal to 1, or larger than 1, respectively.

Omega can be measured. According to the best current measurements, the critical density of the universe, as determined by the rate of expansion, is equivalent to a few atoms of hydrogen in a box a meter on a side, roughly the density achieved by spreading the mass of a postage stamp through a sphere about the size of the earth. The average density of all the matter we can detect by its radiation or by its gravitational effect is about one-tenth this critical value.

This result, as well as other observations, seems to suggest that our universe is of the open variety.

There are uncertainties in these estimates, mostly connected with uncertainties about cosmic distances. If the universe were precisely homogeneous and uniformly expanding, then the rate of expansion of the universe (the Hubble constant) could be determined by measuring the recessional speed and distance of any galaxy, near or far. Conversely, the distance to any galaxy could be determined from its redshift and the application of Hubble's law. The average density of matter could be figured by estimating the mass of a group of galaxies and then dividing by the volume of space that it occupies. However, the universe is not at all homogeneous; structure is apparent on every scale we have studied. Because of local inhomogeneities, the rate of expansion of the universe and the average density of matter need to be measured over as large a region as possible. Accurate determinations of distances to galaxies are needed for both of these measurements.

One possible means of accurately determining distance involves the scattering of the cosmic background radiation by hot gas in clusters of galaxies by a process called the Sunyaev-Zeldovich effect. The hot gas gives a slight energy boost to the radio waves as they pass through it on their way to the earth. As a result of measuring both the change in energy of the radio waves and the x-ray emission from the hot gas, the distance to the cluster of galaxies will be well determined. With the MMA, AXAF, and other instruments, astronomers hope to make these measurements in the coming decade. Such measurements repeated for a large number of galaxy clusters would permit a more accurate determination of the rate of expansion of the universe.

Likewise, studies of the velocities and distances to a large number of galaxies using new 4-m telescopes equipped with spectrographs of novel design are capable of measuring the distances to hundreds of galaxies at a time. These data could pin down the local values of both omega and the Hubble constant. Peculiar velocities of galaxies depend on the extra amount of matter concentrated in a region, over and above the average density of cosmic matter. Measurements of peculiar velocities, together with a knowledge of how much matter there is above the average, lead to an estimate for omega.

Despite these difficulties, cosmologists are almost positive that the value of omega lies between 0.1 and 2. Enough matter has been identified so that omega cannot be less than about 0.1. On the upper end, an omega larger than 2, together with the current rate of expansion, would translate to an age of the universe that is less than the age of the earth as determined by radioactive dating.

The inflationary universe model firmly predicts that omega should be equal to 1, exactly. On this basis, the model can in principle be either ruled out or supported from observational evidence. Since current observations suggest a value of omega closer to 0.1, scientists who believe on theoretical grounds that the model is right must have faith that an enormous amount of mass is hiding from us, escaping detection, perhaps in a uniform and tenuous gas of particles between galaxies.

To summarize, the light-emitting matter we see accounts for sufficient mass to make omega about 0.01; the unseen but gravitationally detected dark matter accounts for another factor of 10 of mass, increasing omega to about 0.1. Advocates of the inflationary universe model must hypothesize that space contains yet 10 times more mass—not only unseen but undetected and composed of some exotic species of matter. As mentioned earlier, some candidate particles will be searched for in the laboratory using "dark matter" detectors currently under development.

Some of the general features of the inflationary universe model are so appealing that many astronomers and physicists believe that some form of the idea is correct. It is sobering to realize that the highly influential inflationary universe model was unknown only a decade ago. Like Supernova 1987A, the idea exploded. We should expect similar observational and theoretical surprises in the future.

Even if the cosmos is infinite in extent, only a limited volume is visible to us at any moment: we can see only as far as light has traveled since the Big Bang. As we look farther into space, we are seeing light that has traveled longer to reach us. Eventually, at some distance, the light just now reaching our telescopes was emitted at the moment of the Big Bang. That distance marks the edge of the currently observable universe, some 10 billion to 20 billion light-years away. We cannot see farther because there has not been time for light to have traveled from there to here. And we have no way of knowing what lies beyond that edge. Some theorists have recently proposed that extremely distant regions might have different properties from the cosmos we know—different forces, different types of particles, even different dimensionalities of space. In such a universe, it would be impossible for us ever to learn about more than a tiny fraction of the possibilities and realities of nature.

Even in the universe we can see, many fundamental surprises surely await us. It is likely that major properties of the universe are yet unknown. The expansion of the universe was unknown in 1920 and the existence of quasars unsuspected in 1960. Who can imagine what astronomers will find by the year 2000?

3

Existing Programs

INTRODUCTION

Astronomers observe the universe in unique ways from space, from the ground, and even from underground. The variety of astronomical objects and the number of instrumental techniques are remarkable. The committee summarizes in this chapter the most important instrumental projects that are operating or under way. The committee also describes programs in theoretical and laboratory astrophysics and in particle astrophysics that are important to the research base in astronomy. In the discussion below, the committee highlights planned enhancements to certain ongoing programs that would be particularly effective scientifically. The committee emphasizes that a vigorous research program requires grants to individual astronomers and broad access to telescopes; these issues are addressed in Chapters 1 and 7. For any of the programs discussed in this chapter or the next ("New Initiatives") to be effective, there must be adequate support for data analysis, interpretation, and theoretical studies.

GROUND-BASED ASTRONOMY

Optical and Infrared Astronomy

Ground-based observations in optical and infrared regions of the spectrum are central to our understanding of astronomical objects and the physical processes that shape their evolution. In the past decade, ground-based observations have changed the scientist's paradigm of the universe from that of an almost boringly uniform expansion to one of startling inhomogeneity that requires huge amounts of unseen matter. Technological breakthroughs in fabricating large mirrors of superb optical quality and in correcting for the distortions introduced by turbulence in the earth's atmosphere now make it possible to build new facilities with unprecedented sensitivity and spatial resolution.

The recommendations for new equipment (Chapter 1) are designed to make a new generation of powerful facilities available to the U.S. astronomical community. The specific recommendations for investment at the federal level reflect the assumption that, in addition, private and state funds will enable ongoing collaborations to complete one or more 8- to 10-m telescopes in the 1990s. The Keck 10-m telescope, the largest optical telescope in the world, developed by the California Institute of Technology and the University of California, is already under construction on Mauna Kea. The Smithsonian Institution and the University of Arizona are converting the Multiple-Mirror Telescope in Arizona to use a single 6.5-m mirror. Other private consortia are designing additional large telescopes. Pennsylvania State University and the University of Texas have designed a segmented 8-m telescope suitable for spectroscopic surveys. The Columbus project, consisting of two 8-m telescopes on a single mount, is being planned by the University of Arizona, Ohio State University, and a foreign partner, Italy. The Magellan project, currently under consideration by the Carnegie Institution, the University of Arizona, and Johns Hopkins University, aims to put a single 8-m telescope at Las Campanas, Chile (Table 3.1).

Although, under current guidelines, little or no observing time on these large new telescopes will be available to the general astronomical community in the United States, the private telescopes, including the Keck, Columbus, Magellan, and the Spectroscopic Survey telescopes, are important to a balanced program of astronomical research and greatly augment the national capability. The scientific problems described in Chapter 2 require more observing time on powerful telescopes than two national instruments can provide. It is the combination of the federal investment in large telescopes recommended in Chapter 1 with the private initiatives, described only briefly here, that will assure all-sky access for the U.S. astronomical community. Increased investment in instrumentation at all levels—federal, state, and private—will be required to make full use of the capabilities of these telescopes.

Several private-federal partnerships to build new 4-m-class telescopes have been formed; the committee views these partnerships as an innovative way to make the best use of limited federal funds. The WIYN (University of Wisconsin, Indiana University, Yale University, and the National Optical Astronomy Observatories) telescope project will put a 3.5-m telescope funded by the three universities on Kitt Peak. Operating expenses will be shared by the National Optical Astronomy Observatories (NOAO) and the universities; the national

TABLE 3.1 Large Nonfederal Telescope Projects

Project	Location	No. of Telescopes × Size	Status
Keck	U.S. (Caltech, Univ. of California)	1 × 10 m	Construction
Very Large Telescope	Europe (ESO)	4 × 8 m	Funded
MMT conversion	U.S. (Smithsonian Inst., Univ. of Arizona)	1 × 6.5 m	Funded
Spectroscopic Survey Telescope	U.S. (Penn State Univ., Univ. of Texas)	1 × 8 m	Proposed
Columbus	U.S. (Ohio State Univ., Univ. of Arizona), Italy	2 × 8 m	Proposed
Japan National Telescope	Japan	1 × 7 m	Proposed
Keck II	U.S. (Caltech, Univ. of California, NASA)	1 × 10 m	Proposed
Magellan	U.S. (Carnegie Inst., Univ. of Arizona, Johns Hopkins Univ.)	1 × 8 m	Proposed

Note: Access to all these telescopes is restricted largely by institutional affiliation or nationality.

community will have access to 40 percent of the telescope time. The 3.5-m ARC telescope is being built by a consortium of universities (University of Chicago, University of Washington, Princeton University, New Mexico State University, and Washington State University). The NSF provided a fraction of the capital funding for ARC with the stipulation that about 10 to 15 percent of the time on the telescope would be available for users outside the consortium. Harvard University and Cambridge University are planning to build a 4-m telescope in the Southern Hemisphere.

LARGE MIRRORS

The Steward Observatory Mirror Laboratory at the University of Arizona operates at the forefront of optical technology and constitutes a crucial element in the U.S. astronomy program. The laboratory is developing advanced techniques of optical design, casting, polishing, and testing, which are necessary to make the 8-m-diameter mirrors needed by the next generation of telescopes (Plate 3.1). The laboratory also plays a unique role in training students in optical engineering and in strengthening interactions between industry, government, and the university community.

ADAPTIVE OPTICS AND INTERFEROMETRY

The implementation of new technologies has permitted astronomers to achieve enhanced spatial resolution from the ground at progressively shorter wavelengths, first in the radio region of the spectrum and, during the coming decade, at infrared and optical wavelengths. Recently, European astronomers have demonstrated impressively that adaptive optics can be successfully applied to astronomy. Adaptive optics permits diffraction-limited imaging over the full aperture of large telescopes. The pioneering U.S. effort in this area was cut back because of budget reductions at the NSF. The committee believes this program should be renewed so that the United States can take a leadership role in adaptive optics. Hence the committee has established adaptive optics as its highest-priority, moderate ground-based program (Chapter 1).

As discussed in more detail in Chapter 4, optical and infrared interferometry can achieve high spatial resolution over baselines longer than the aperture of a single mirror. The technology needed for the application of optical interferometry to faint sources was carried out on some bright stars in the 1980s. The Department of the Navy supported the building of two interferometers on Mt. Wilson. The NSF also established a number of modest programs. These demonstration projects deserve continued support.

Radio Astronomy

CENTIMETER WAVELENGTH ASTRONOMY

The Very Large Array (VLA), recommended by the Greenstein Committee (NRC, 1972), operated with great power in the 1980s, annually providing observations for about 600 astronomers and data for over 200 research papers. During the 1980s, new receivers and computing techniques enhanced the power of the VLA to more than 10 times that of the original instrument, all without major changes in telescope hardware. However, the declining NSF budget has caused major problems at the VLA and its parent National Radio Astronomy Observatory (NRAO), as discussed by the Radio Astronomy Panel in the *Working Papers* (NRC, 1991) and summarized briefly here. A decade of diminishing funds has led to deferred maintenance that directly affects the reliability of the array. For example, the railroad track system over which the antennas are moved has deteriorated (Figure 3.1), making it difficult to configure the VLA for observations at different angular resolutions. Some of the vital instrumentation of the VLA is out of date: needed are low-noise receivers, fiber-optic transmission lines, a broadband digital correlator, and more powerful computers for data analysis. These items could improve the sensitivity of the instrument by up to a factor of 10, improve the frequency coverage and spectral resolution, and increase the maximum image size.

FIGURE 3.1 A decade of deferred maintenance and refurbishment has led to a variety of problems at the national observatories, including the deterioration of the railroad tracks used at the VLA to reconfigure the array for operation at different spatial resolutions. Courtesy of the National Radio Astronomy Observatory/Associated Universities, Inc.

The Very Long Baseline Array (VLBA), the highest-priority ground-based initiative recommended by the Field Committee (NRC, 1982), has been funded by NSF and will begin operating in 1992. The VLBA will provide detailed maps of the cores of active galaxies and quasars with sub-milliarcsecond angular resolution and will determine distances to objects in our own and other galaxies from measurements of H_2O masers. The committee is confident that this next step in the application of very long baseline interferometry (VLBI) techniques will return exciting results.

The most spectacular closing of an old facility was the unanticipated collapse of the Green Bank 300-ft telescope. The replacement for the 300-ft telescope will be the Green Bank Telescope, a fully steerable instrument of comparable diameter. The telescope will incorporate novel design features, such as an active surface, that may eventually permit operation at wavelengths as short as 3 mm. The telescope will begin operation in 1995, initially at centimeter and longer wavelengths, for the study of pulsars, active galaxies, and 21-cm hydrogen emission in our own and in distant galaxies. The upgrade of the 1,000-ft Arecibo telescope will improve the sensitivity of that instrument by a factor of 3 to 4 for nonradar observations. Because of its large collecting area, the Arecibo telescope will continue to play a critical role in pulsar studies and

TABLE 3.2 Millimeter and Submillimeter Telescope Projects

Project	Location	No. of Telescopes × Size	Status
Single Dishes			
Nobeyama	Japan	45 m	Operational
IRAM	Europe	30 m	Operational
Maxwell Observatory (JCMT)	U.K.	15 m	Operational
Five Colleges	U.S.	14 m	Operational
NRAO	U.S.	12 m[a]	Operational
CSO	U.S.	10 m[b]	Operational
Bell Laboratories	U.S.	7 m	Operational[c]
Submillimeter Telescope	U.S., Germany	10 m	Funded
Interferometers			
Nobeyama	Japan	6 × 10 m	Operational
Owens Valley	U.S.	3→4 × 10 m[b]	Operational
Berkeley-Illinois-Maryland	U.S.	6→9 × 6 m[b]	Operational
IRAM	Europe	3 × 10 m	Operational
Smithsonian submillimeter array	U.S.	6 × 6 m	Funded
NRAO Millimeter Array	U.S.	40 × 8 m[a]	Proposed

[a]Telescopes without restrictions on access by U.S. astronomers.
[b]Approximately 30 to 50 percent of time available to U.S. community.
[c]To be decommissioned in 1991-1992.

in extragalactic surveys of HI in galaxies even after the Green Bank Telescope is built.

MILLIMETER AND SUBMILLIMETER WAVELENGTH ASTRONOMY

The focus of U.S. activity in millimeter and submillimeter astronomy (Table 3.2) shifted during the 1980s from single dishes to interferometers, beginning with the inauguration of millimeter interferometers at Owens Valley (Caltech) and Hat Creek (Berkeley, University of Illinois, and the University of Maryland). These small arrays of 6- to 10-m telescopes have made important contributions to many fields of galactic and extragalactic astronomy. Important results include, for example, the discovery of molecular gas concentrated at the center of infrared-luminous galaxies and the imaging of protoplanetary disks associated with young stars.

Submillimeter astronomy is expected to become a major field of research in the 1990s. The Caltech Submillimeter Observatory began operations in 1990 with significant support from the NSF and with 50 percent of the observing time available to the national community. This telescope, a small program recommended by the Field Committee, has taken advantage of sensitive receivers and clever telescope fabrication techniques to pioneer observations in the submillimeter wavelength band. Important new initiatives are under way with the forthcoming construction of the Smithsonian Institution's submillimeter array and the University of Arizona-Germany Submillimeter Telescope. The Smithsonian array will make subarcsecond images with high spectral resolution for the first time in this important wavelength band. The NSF is supporting construction of a 1.7-m telescope to investigate the feasibility of conducting submillimeter observations from the South Pole.

Planetary Astronomy

NASA's Infrared Telescope Facility (IRTF) and the Kuiper Airborne Observatory (KAO) continue to provide important new results in planetary astrophysics, including the direct detection of water vapor in Halley's Comet, the imaging of volcanoes on Io, and the detection of numerous molecules, including water, phosphine, and germane, in the atmosphere of Jupiter.

Radar results play a critical role in planetary astronomy. The radar sensitivity of the Arecibo telescope will be increased by a factor of 10 or more in a joint undertaking by NASA and the NSF. The enhanced telescope will make high-spatial-resolution images of many asteroids and comets and probe the subsurface properties of many of the natural satellites such as Phobos, Deimos, Io, and Titan. The committee commends NASA and the NSF for collaborations on the Arecibo telescope.

Solar Astronomy

The sun can be studied at a level of detail that is impossible to achieve for any other star, establishing the foundation of our understanding of all stars. The physical regions inside the sun, especially in the convective zone and below, can be probed by observing the oscillation (or "ringing") of the sun's surface. The technique is similar to studying the interior of the earth by observing seismological waves on the earth's surface. The Global Oscillations Network Group (GONG) project is an international collaboration, supported in this country by the NSF, to set up a chain of telescopes around the world to monitor the sun's oscillations continuously. With GONG data, it will be possible to constrain the interior temperature and density structure of the sun, and to infer its differential rotation as a function of radius, latitude, and depth. The spaceborne complement of GONG is a NASA-funded helioseismology instrument on the European Space Agency's (ESA's) Solar-Heliospheric Observatory (SOHO) mission.

- **The committee regards the completion of the GONG network, with adequate support for its continued operation and data analysis, as being of fundamental importance.**

The facilities of the National Solar Observatory (NSO) are critically important to the solar community, both because few university astronomy departments maintain solar facilities and because NSO telescopes are among the best in the world. Yet budgetary pressures, accumulated over many years, have weakened the NSO. For example, the observatory has been unable to purchase state-of-the-art infrared or visible-light detector arrays to support solar adaptive-optics experiments at an appropriate pace. This situation should be corrected and is one reason that this committee's highest priority for ground-based astronomy is renewed investment in the infrastructure.

Solar neutrino experiments constitute a complementary way of looking inside the sun. The nuclear fusion reactions that cause the sun to shine occur deep within the sun's core but are revealed directly by observations of particles called neutrinos. The United States operates the chlorine solar neutrino experiment, which detects rare high-energy neutrinos of the electron type. The United States also collaborates with the Soviet Union, Germany, France, Italy, and Israel on experiments using gallium detectors to detect electron neutrinos from the basic proton-proton reaction and with Canada and the United Kingdom to construct a detector of heavy water to observe higher-energy neutrinos of all types. In addition, the United States has a potentially important collaboration with Italy and the Soviet Union to observe the beryllium neutrino line with a liquid scintillator. These observatories have complementary functions, including the study of variations of the neutrino flux with phase in the solar cycle.

The Search for Extraterrestrial Intelligence

Ours is the first generation that can realistically hope to detect signals from another civilization in the galaxy. The search for extraterrestrial intelligence (SETI), involves, in part, astronomical techniques and is endorsed by the committee as a significant scientific enterprise. Indeed, the discovery in the last decade of planetary disks, and the continuing discovery of highly complex organic molecules in the interstellar medium, lend even greater scientific support to this enterprise. Discovery of intelligent life beyond the earth would have profound effects for all humanity. NASA's decade-long Microwave Observing Program is based on a particular set of assumptions and techniques for exploring the SETI problem. This committee, like the Field Committee before it, believes strongly that the speculative nature of the subject also demands continued development of innovative technology and algorithms. A strong peer-reviewed, university-based program should be an integral part of this effort.

SPACE ASTRONOMY

During the 1980s, the initial reliance on the Space Shuttle for access to space, followed by the Challenger disaster, slowed the rate of progress in U.S.

space science. In contrast to the fruitful 1970s, only two American astronomical satellites were launched in the 1980s, and leadership in areas the United States had pioneered, such as x-ray astronomy, moved to Europe, the Soviet Union, and Japan. The currently planned program in space astronomy, described in the *Strategic Plan* (NASA, 1988, 1989) for NASA's Office of Space Science and Applications, can reverse this trend. NASA's plan includes telescopes that range from small payloads like the Submillimeter Wave Astronomy Satellite to be launched in 1995, to the four Great Observatories (discussed below). This section describes the ongoing program in space astronomy that underpins the recommendations for new projects made in Chapter 1.

The Great Observatories

The first Great Observatory, the Hubble Space Telescope (HST), was launched in April 1990. The second, the Gamma Ray Observatory (GRO), is currently scheduled for launch in 1991, after the publication of this report. Construction of the third, the Advanced X-ray Astrophysics Facility (AXAF), the highest-priority major new program of the Field Committee in 1982, is under way. As discussed in Chapter 1, this committee endorses the plan to complete the Great Observatories, which are essential for reaching the frontiers of the universe. The Space Infrared Telescope Facility (SIRTF) is the only Great Observatory awaiting initiation and is this committee's highest-priority large equipment initiative for the decade of the 1990s.

HUBBLE SPACE TELESCOPE

A manufacturing flaw in the primary mirror of the HST will prevent it from forming images of faint objects with a resolution greater than about 1 arcsecond, until after the installation of a second generation of instruments with correcting optics. But important observations will be possible with HST even with reduced performance. The General Observer program for using the HST observatory, which is conducted by the Space Telescope Science Institute, makes possible the use of HST's frontier astronomical facilities by U.S. and foreign scientists. The Hubble Fellowship Program for recent PhDs helps to train some of the best young researchers in space astronomy.

At this stage, it appears likely that close to the full potential of HST can be achieved by installing either new instruments or corrective optical elements, although there is no guarantee that all technical and practical problems will be overcome. A replacement, with corrective optics, for the general-purpose Wide-Field/Planetary Camera (WF/PC) is under construction and is scheduled for installation in 1993; the replacement camera is essential to carry out the fundamental goals of this observatory.

In addition to the improved camera, NASA selected in 1988 two other new instruments, the Space Telescope Imaging Spectrometer (STIS) and the Near-Infrared Camera and Multi-Objective Spectrometer (NICMOS). Both of these instruments can also be fitted with appropriate optics to remove the aberrations in the HST images and so achieve the high spatial resolution possible with an orbiting telescope. These new instruments will greatly enhance the power of the HST observatory. STIS will increase the speed with which some critical ultraviolet and visible-light observations can be made by a factor as large as 100 or more and will make possible spatially resolved spectroscopy. NICMOS will use two-dimensional arrays for imaging faint, complex fields with 0.1-arcsecond spatial resolution, and for spectroscopy with frequency resolution up to 10,000 at near-infrared wavelengths. In this wavelength range, the background radiation affecting HST is 100 times smaller than that affecting terrestrial telescopes, which must observe through the earth's time-variable atmosphere.

- **The committee considers the prompt installation of the new WF/PC to be of critical importance to space astronomy. Corrective optics may restore most of the capabilities of the other instruments. All three second-generation instruments (WF/PC, STIS, and NICMOS) must be installed and work well in order for HST to attain its full scientific goals.**

GAMMA RAY OBSERVATORY

The Gamma Ray Observatory will study a broad range of topics, including accretion processes around neutron stars, the origin of gamma-ray bursts, nucleosynthesis in supernovae, interactions of cosmic rays with interstellar matter, and energy production by giant black holes in galactic nuclei. GRO's instruments have sensitivities and angular resolutions more than an order of magnitude better than those available on previous missions. The expected value of GRO's dataset mandates a vigorous, peer-reviewed program of investigations by the broad astronomical community.

ADVANCED X-RAY ASTROPHYSICS FACILITY

The Advanced X-ray Astrophysics Facility, the number-one-priority major new program recommended by the Field Committee in 1982, will return the United States to preeminence in x-ray astronomy, a field pioneered by NASA's earliest x-ray-detecting sounding rockets and satellites. Construction of AXAF is under way, with a launch planned for the latter part of the 1990s. AXAF will have a major impact on almost all areas of astronomy, including studies of the coronae of nearby stars, mapping of energetic galaxies, and detection of hot gas within distant clusters of galaxies. AXAF has a strong technical and scientific heritage from previous x-ray missions; in many ways, it is a scaled-up version of the successful Einstein Observatory launched more than a decade ago. This heritage, together with management from a single NASA center, designation of a prime contractor, project involvement of experienced NASA and non-NASA scientists, and scheduled end-to-end testing, means that AXAF should avoid many of the problems that plagued the HST program.

- **The committee reaffirms the Field Committee decision that made AXAF the highest-priority large program of the 1980s and stresses the importance to all astronomy of deploying AXAF as soon as possible. The committee expresses its strong support for the early establishment of a science center that would help NASA maximize the scientific return from AXAF.**

The Explorer Program

The Explorer queue for astronomy and astrophysics currently includes five missions (Table 3.3) that, according to current NASA plans, will be completed before new projects in either the Delta class or the Small Explorer (SMEX) class can be flown. The currently planned flight rate for astrophysics is one Delta-class Explorer and one SMEX-class Explorer approximately every two years. As discussed in Chapters 1 and 7, the committee has recommended acceleration of the Delta-class and SMEX-class Explorer programs.

Two Explorer missions will operate at ultraviolet wavelengths shorter than HST's limit of 120 nm. The Extreme Ultraviolet Explorer (EUVE) will carry out an all-sky survey in several bands covering wavelengths between 7 and 76 nm, extending the ultraviolet survey of the Roentgen Satellite (ROSAT) to longer wavelengths. Spectroscopic observations with a resolving power of 250 will be carried out on EUVE through a guest observer program. A deep survey will cover about 1 percent of the sky with a sensitivity at least 10 times greater than that of the all-sky survey and will provide guest investigators the opportunity to make spectroscopic observations of interesting new objects. EUVE will add greatly to our understanding of hot, young white dwarfs, cataclysmic variables, stellar coronae, and the local interstellar medium. The Far Ultraviolet Spectroscopy Explorer (FUSE) will make spectroscopic observations shortward of 120 nm with unprecedented sensitivity (100,000 times greater than the sensitivity of the Copernicus telescope launched 20 years ago) and will bridge the spectral gap between HST and AXAF. FUSE will open a window that contains the fundamental resonance transitions of atomic and molecular species that can be used to probe physical processes in the early universe, measure the abundance of deuterium in a variety of environments, and determine the physical conditions in the interstellar medium in distant and evolving galaxies. As discussed in Chapters 1 and 4, the highest-priority, moderate space-based initiative is the acquisition of an independent, dedicated spacecraft for FUSE.

The X-ray Timing Explorer (XTE) will make spectroscopic and photometric

TABLE 3.3 Currently Funded Explorers

Program	Date Selected	Planned Launch Date[a]
Delta-Class Explorers		
Extreme Ultraviolet Explorer (EUVE)	1982	1992
X-ray Timing Explorer (XTE)	1984	1996
Advanced Composition Explorer (ACE)	1989	1997
Far Ultraviolet Spectroscopy Explorer (FUSE)	1989	1999
Small Explorers (SMEX)		
Submillimeter Wave Astronomy Satellite (SWAS)	1989	1995

[a]Launch dates after 1995 are uncertain.

observations in the 1- to 100-keV range with microsecond temporal resolution. XTE will advance our understanding of the physics of accretion flows around neutron stars and black holes, and of relativistic plasmas in the nuclei of active galaxies.

The Advanced Composition Explorer (ACE) will study the isotopic and elemental abundances of cosmic rays over a broad range of energies.

The Small Explorer program was initiated in the late 1980s to provide rapid access to space for payloads weighing less than about 200 kg. The stringent requirements of astronomy instruments often require pointing systems that weigh almost this amount by themselves; nevertheless, a number of imaginative proposals have been made. The Submillimeter Wave Astronomy Satellite has been selected as the first space mission to explore this wavelength band. As enhanced launch vehicles become available, a broader range of astronomical projects will become possible.

NASA is to be commended for its support of the analysis of data from the Infrared Astronomical Satellite (IRAS). The success of IRAS and of its General Investigator program serve as a model for the active support of Explorer missions. The committee urges NASA to provide strong support for the analysis of data from the Cosmic Background Explorer (COBE), including a vigorous guest investigator program.

The Suborbital Program

NASA's suborbital program trains students, tests instruments, and explores new scientific ideas by flying telescopes in rockets, balloons, and aircraft. This activity is NASA's only space science hardware program that operates on the time scale of a graduate student's career, allowing students and postdoctoral associates to be involved in all aspects of developing new instruments. Important scientific results, including the discovery of celestial x-ray sources, the measurement of the dipole anisotropy of the microwave background, discovery of variable gamma-ray emission from the galactic center, and critical infrared and gamma-ray measurements of Supernova 1987A, have come from the suborbital program.

The Kuiper Airborne Observatory (KAO), a 0.9-m telescope operating for 15 years in a C-141 aircraft, opened up far-infrared and submillimeter wavelengths to scientific investigation, produced over 700 scientific papers, and trained 40 PhD students. The importance of the suborbital program for the training of instrumentalists is exemplified by the fact that 80 percent of the U.S. science team on the successful IRAS program had worked previously on the KAO. As discussed in Chapters 1 and 4, one of the major recommendations of this report is to fund the Stratospheric Observatory for Far-Infrared Astronomy (SOFIA) as a successor to the KAO.

International Collaborations

Astronomy has traditionally been an international enterprise, and as space missions have become more complex, collaborations with foreign colleagues have made possible important programs that otherwise would have been unaffordable. NASA has used resources from the Explorer program to support U.S. scientists to fly instruments on foreign spacecraft in the absence of flight opportunities on American spacecraft. The ROSAT x-ray telescope was launched in 1990 as a collaboration between Germany, the United Kingdom, and the United States. ROSAT's high-sensitivity survey will produce an all-sky catalog of more than 100,000 galactic and extragalactic x-ray sources. Following the survey, U.S. and European researchers will carry out pointed observations of many known and newly discovered objects. U.S. instruments are also planned for the Japanese-U.S. ASTRO-D mission, the Soviet-French Spectrum X-Gamma, and the ESA's X-ray Multi-Mirror Mission. Other important missions include radio interferometry from space with the Japanese VSOP and the Soviet RadioAstron missions and a proposed NASA-ESA collaboration on an orbiting submillimeter telescope. Participation by U.S. scientists in ESA's Infrared Space Observatory (ISO) will provide valuable astrophysical data and help define SIRTF's scientific program. In Chapter 1, the committee recommends a new budgetary line to cover the costs of international collaborations carrying U.S. instruments, an activity that currently is supported out of the Explorer program.

Shuttle Payloads

After the Challenger accident, NASA revived the mixed-fleet philosophy that utilizes unmanned boosters, like the Deltas. The committee strongly endorses this strategy of making unmanned boosters available as the main launch vehicles for new astronomy missions. The committee recognizes that some important missions, such as the ASTRO telescope, the HST second-generation instruments, and the High-Energy Transient Experiment (HETE), are dependent on the Shuttle for either launching or servicing. While this report was in its final stages of preparation, ASTRO had a successful nine-day mission producing x-ray, extreme-UV, and UV spectra; UV polarization measurements; and UV images of a wide range of astronomical objects. A preliminary inspection of those data suggests that many discoveries will follow from the full analysis of the dataset.

Technology Development

An imaginative, innovative program of technology development is a prerequisite for the missions of the next century. Current programs such as the NICMOS instrument for HST and the entire SIRTF mission benefited from NASA's investment in advanced detector technologies. AXAF benefited from the application of bolometers (developed in infrared astronomy) and charge-coupled devices (CCDs; developed in optical astronomy) to x-ray astronomy. Future submillimeter and far-infrared telescopes as large as 10 m may use technology being developed by NASA for lightweight optics with surface accuracies around 1 μm.

Technology development set in motion long before critical mission milestones is cost-effective. The committee concurs with the strong support given by the Committee on Space Policy (the "Stever Report"; NAS-NAE, 1988) for vigorous technology development across NASA's entire space astronomy program. Chapter 1 recommends new technology initiatives in support of a generation of telescopes beyond the Great Observatories. As discussed in Chapter 6, "Astronomy from the Moon," the committee strongly favors phased technology-development efforts that progress from laboratory test beds, to modest instruments and precursor missions with significant scientific goals, and finally, to large sophisticated observatories.

THEORETICAL AND LABORATORY ASTROPHYSICS

Theory provides the paradigms within which observations are planned and interpreted, and it must also respond to unexpected observational discoveries. A strong theoretical community also makes motivating, and often surprising, predictions about what might be seen. This predictive capability is often crucial in designing new instruments: for example, the characteristics of the Cosmic Background Imager recommended in Chapter 1 are determined by theoretical calculations of fluctuations in the microwave background, and the proposed gamma-ray-spectroscopy Explorer mission is designed to detect theoretically predicted lines from supernovae in other galaxies. Solar neutrino experiments

are meaningful only in relation to theoretical predictions; a few radioactive decays, more or less, in a thousand tons of detector material are significant only in comparison with predicted rates.

Theory and observation together provide a coherent view of the universe. The success of modern astrophysics illustrates the close interdependence of theory, observation, and experiment. As described in the report from the Theory and Laboratory Astrophysics Panel in the *Working Papers* (NRC, 1991), the vitality of astronomy and astrophysics requires strong support for theoretical investigations as well as for experimental and observational programs.

NASA responded positively to the Field Committee's recommendation that it establish a strong, broad program in theoretical astrophysics. Many of the arguments made by the Field Committee in 1982 still apply. New observations from telescopes operating in the 1990s are likely to "consume" a great deal of existing theory and point the way to more sophisticated modeling and interpretation. If the full scientific benefits are to be realized for existing and planned space missions, then theoretical interpretations are required.

- **The committee believes that NASA's support for theoretical investigations should grow in approximate proportion to NASA's support for the analysis and interpretation of observational data.**

Grants to individual investigators have traditionally been one of the strengths of U.S. science, particularly at NSF. Theory is by its nature often both multidisciplinary and interdisciplinary, and therefore it does not readily fit into the object-oriented classification of the Astronomy Division at NSF. The existing organization of the NSF's grants program is damaging to theoretical investigations, which most often are not tied to a specific project or subject area.

- **The committee recommends that the NSF establish a separate, adequately funded theory program in the Astronomy Division. Theory is frequently interdisciplinary and therefore is not easily categorized within the existing object-oriented classification.**

The interpretation of results from ground- and space-based telescopes requires knowledge of nuclear, atomic, and molecular physics, as well as of the properties of interstellar dust and solid surfaces. The present level of support is inadequate to provide observers and theoreticians with the data they need to interpret, for example, the intensity in a particular spectral line in terms of the physical characteristics of the emitting plasma. It is important that NASA and NSF support the relatively inexpensive laboratory and theoretical work that is important to their astrophysics initiatives and is crucial to the interpretation of the results from their major observatories.

The Department of Energy (DOE) has traditionally supported theoretical research in areas of importance and immediate applicability to astronomy and astrophysics, including plasma physics, atomic physics, nuclear physics, radiative transfer, properties of matter, cosmology, and the physics of the early universe. The opportunity is great for fruitful cross-fertilization between astrophysics and the laboratory-oriented physics that is supported by the DOE.

PARTICLE ASTROPHYSICS

The study of solar neutrinos contributes to our understanding of stellar interiors and of fundamental physics. The discrepancy between the predicted rates of solar neutrino production and the values measured in the Homestake mine since the 1960s, and recently confirmed by the Japanese Kamiokande II experiment, has raised important questions for both astrophysics and particle physics. The first preliminary answers to these questions will become available in the 1990s from the Soviet-American and Western European-Israeli-American experiments using gallium detectors.

The important new experimental results are being obtained using neutrino observatories in other countries, including Canada (Sudbury), Italy (Gran Sasso), Japan (Kamiokande), and the Soviet Union (SAGE), although many of the experimental techniques and theoretical ideas were developed in the United States. However, the international collaborations are strong, and some involve talented American scientists working at the frontiers of research. The committee recommends in Chapter 1 the development of the technology for a new generation of U.S. solar neutrino experiments (see Table 1.3). These detectors would also be sensitive to supernova neutrinos.

- **The committee urges the DOE and NSF to continue to support American participation in international solar neutrino experiments.**

The serendipitous detection of neutrinos from Supernova 1987A confirmed the basic ideas of stellar collapse, including the order of magnitude of the total neutrino energy emitted, the time scale for the neutrinos to escape, and the characteristic energy of an individual neutrino. Only about 20 neutrinos were detected, but the observation validated theoretical insights that originated more than half a century ago. More diagnostic measurements can be made of future supernovae with detectors designed specifically for this purpose.

Two fundamental programs in particle astrophysics involve the detection of particles with very high energies: gamma rays with energies in the range 10^{11} to 10^{14} eV and cosmic rays up to 10^{20} eV. The Whipple Observatory of the Smithsonian Institution has detected gamma rays above 3×10^{11} eV from the Crab Nebula, and there are possible detections from other observatories of even higher energies from x-ray binary stars. The "Granite" telescope, currently under construction at Mt. Hopkins, will have improved sensitivity to radiation with energies up to 10^{12} eV. Upgraded airshower facilities at Los

Alamos, New Mexico, and Dugway, Utah, offer the hope of confirming the detection of 10^{14}-eV gamma rays and extending the observations to new objects. The development of new gamma-ray airshower detectors is recommended in Chapter 1 (see Table 1.3).

The existing Fly's Eye telescope has convincingly measured cosmic rays with energies greater than 10^{19} eV, whose origin is still a puzzle. The new Fly's Eye telescope recommended by the committee in Chapter 1 represents an outgrowth of this exciting work. These ongoing programs, which lay the groundwork for more ambitious projects that may be required later in the decade, should be pursued vigorously.

The 1980s saw many interactions between astronomers and physicists on important theoretical questions, such as the origin of the predominance of matter over antimatter and the evolution of large-scale structure in the universe. Only in the first few microseconds after the "Big Bang" were the conditions of density and temperature extreme enough to produce some of the reactions predicted by modern theories of elementary particle physics. Some particle physics models can be tested by comparing their predictions for cosmology and for elemental abundances with astronomical observations.

One of the most striking examples of these interactions involves the existence and nature of "dark matter." As discussed in Chapter 2, a feature of much of the recent work connecting cosmology and particle physics is the requirement that the universe contains much more matter than is seen. In fact, for many years astronomers have quite independently been obtaining observations of galaxies and clusters of galaxies that suggest that as much as 90 percent of the matter inferred to be present from its gravitational effects has not been seen. Many of the explanations involve exotic particles such as massive neutrinos, axions, or weakly interacting massive particles. Imaginative experiments have been proposed to detect contributions to the missing mass from various particles. The committee has recommended (see Table 1.3) that the technology for various dark matter detectors be developed in the 1990s.

4

New Initiatives

INTRODUCTION

Progress in understanding the universe comes from new or improved observing techniques and from theoretical insights. Over the next decade astronomers will use new technologies to increase dramatically both the sensitivity and the spatial and spectral resolution of their observations. The program of new initiatives recommended in Chapter 1 (see Table 1.1) calls for telescopes that improve these key areas of performance by factors of tens to millions. Chapter 1 contains thumbnail sketches of the major initiatives. This chapter gives more complete technical descriptions of the large and moderate programs and their anticipated scientific returns. Most of the instrumental initiatives result from advances in infrared technology, in spatial resolution, in the construction of large telescopes, and in the linking of electronic detectors to powerful computers.

Four areas of research are currently of particular significance, and these will be used to illustrate the importance of the recommended instruments. They are the following:

- **The birth of stars and planets.** Observations and theory suggest that the formation of a rotating disk of gas and dust, containing enough material to make planets, is a natural stage in the birth of stars like the sun. Astronomers now strongly suspect that planets form in such protoplanetary disks of gas and dust around young stars. New telescopes with high infrared sensitivity or hundredth-of-an-arcsecond spatial resolution will be used to investigate how

TABLE 4.1 Characteristics of Recommended Large and Moderate Programs

Project	Wavelength Coverage	Increase in Sensitivity[a]	Typical Spatial Resolution (arcsec)	Typical Spectral Resolution
SIRTF	2.5 to 700 μm	30 to 10^3	7.5 $(\lambda/30)^b$	3 to 2000
Infrared-optimized 8-m telescope	1 to 30 μm	3 to 10	0.7 $(\lambda/30)^b$	3 to 10^5
MMA	900 to 7000 μm	30	0.07 $(\lambda/1000)^b$	100 to 10^6
Southern 8-m telescope	0.3 to 2 μm	1	0.1 at 0.5 μm	3 to 10^5
Adaptive optics	0.3 to 10 μm	50	0.1^c	3 to 10^4
SOFIA	1 to 1000 μm	10	2.5 $(\lambda/30)^b$	3 to 10^6
Optical and infrared interferometers	0.3 to 10 μm	10^3	$2.4 \times 10^{-4}\lambda^{b,d}$	3 to 100
AIM	0.1 to 1 μm		$< 30 \times 10^{-6}$	3 to 100
LEST	0.3 to 1 μm		0.1 at 0.5 μm	10 to 10^5
VLA extension	1 to 100 cm		0.1 at 6 cm	50 to 10^6

[a]Relative to other existing or planned facilities of comparable nature.
[b]Wavelength in microns.
[c]2 μm with 8-m telescope.
[d]1-km baseline.

stars form out of interstellar gas and how the disks that surround young stars might evolve into planets.

- **Active galaxies and quasars.** Are active galaxies powered by black holes? What is the link between infrared-luminous galaxies and quasars? These questions will be explored using ground- and space-based telescopes with high spatial and spectral resolution operating at wavelengths from radio to gamma rays.
- **Large-scale structure of the universe.** Observations indicate that clusters of galaxies are strewn in sheets and filaments surrounding large voids and that as much as 90 percent of the mass in the universe may have escaped detection. New ground-based instruments operating at infrared, optical, and radio wavelengths will map the three-dimensional distribution of matter in the universe out to distances of a billion light-years and may reveal the physical processes that create such unexpected patterns.
- **The birth of galaxies.** The greatest single burst of star formation in the history of the universe attended the birth of galaxies between 10 billion and 15 billion years ago. New instruments will be used to investigate how galaxies and quasars come into being and how their existence can be reconciled with models of the infant universe. Searches for protogalaxies will require observations with sensitive, large-format arrays of infrared detectors on telescopes in space and on the ground.

Table 4.1 summarizes the performance of the recommended large and

TABLE 4.2 Contribution of Major Recommended Programs to Scientific Themes

Program	Formation of Planets and Stars	Origin of Energetic Galaxies and Quasars	Distribution of Matter and Galaxies	Formation and Evolution of Galaxies
SIRTF	Comets, primordial solar system Planetary debris disks Protostellar winds Brown dwarf surveys	Spectra of luminous galaxies to $z = 5$	Deep 10- to 100-μm surveys Galactic halos and missing mass	Protogalaxies Galaxy evolution
Infrared-optimized 8-m telescope	Imaging and spectra of protostellar disks and jets Planet searches Planetary structure, atmospheres	High-resolution imaging and spectra of energetic galaxies	Deep 2-μm surveys Galaxy evolution	High-z protogalaxies High-z supernovae
MMA	Motions, structure, chemistry of molecular clouds and protostellar disks Comets Planetary atmospheres	Motions, structure, chemistry of quasars, luminous galaxies	CO redshift surveys	Dusty galaxies at $z \leq 10$ Sunyaev-Zeldovich effect
Southern 8-m telescope	Doppler searches for planets	Spectra of active galaxies and quasars Deep imaging of galaxies	Quasar abosrption lines $z = 1$ redshift surveys	Galaxy evolution Quasar distribution Distant globular clusters

moderate projects in key areas. Table 4.2 compares the contributions of the major recommended projects to the scientific themes. More detailed descriptions of the instruments and the science that they might accomplish can be found in the *Working Papers* (NRC, 1991).

THE DECADE OF THE INFRARED

The infrared and submillimeter portion of the spectrum, from 1 μm to 1000 μm, is poorly explored but is of fundamental importance for almost all aspects of astronomy, from solar system studies to cosmology. Four factors make these wavelengths critical: (1) the expansion of the universe shifts the radiation from primeval objects out of the ultraviolet and visible bands into the infrared; (2) most of the known mass in galaxies is in the form of cool stars that are brightest at wavelengths in the range 1 to 10 μm, depending on the distance to the galaxy; (3) the dust associated with cold gas and star-forming regions obscures objects at wavelengths shorter than 1 μm, but glows brightly at longer wavelengths due to the absorbed energy; and (4) atoms and molecules have rich infrared spectra that can be used to probe the density, temperature, and elemental abundances of astronomical objects.

The technology for detecting infrared and submillimeter radiation has been revolutionized in the last 10 years. A decade ago, astronomers used single detectors on ground-based telescopes to observe in a few spectral windows between 1 and 30 μm. Radio astronomers struggled to detect radiation at wavelengths as short as 1,000 μm. Pioneering astronomers used balloons or airborne telescopes to work between 30 and 1000 μm. In 1983 the Infrared Astronomical Satellite (IRAS) demonstrated that a telescope cooled with liquid helium could approach the theoretical sensitivity limit set by the faint light emitted by interplanetary dust grains. The 1,000-fold increase in sensitivity compared with that of earth-bound telescopes permitted IRAS to survey the entire sky at wavelengths from 12 to 100 μm and to discover important new phenomena, including trails of solid material behind comets, disks of solid material orbiting nearby stars—possibly the remnants of planet formation (Plates 2.1 and 4.1)—and luminous galaxies emitting more than 90 percent of their energy in the infrared. At wavelengths between 100 and 1000 μm, new techniques for detecting radiation using the Kuiper Airborne Observatory (KAO) and ground-based telescopes like the Caltech Submillimeter Observatory (CSO) led to the discovery of spectral lines from atoms and molecules that brought new information about planetary atmospheres, star formation, and interstellar chemistry. Most recently, the development of arrays of detectors operating from 1 to 200 μm has led to the replacement of single-channel photometers by cameras and two-dimensional spectrographs with 50,000 or more individual detectors.

The large initiative accorded the highest priority by this committee is the Space Infrared Telescope Facility (SIRTF), consisting of a 0.9-m telescope cooled by a five-year supply of liquid helium and mounted on a free-flying spacecraft. As proposed, SIRTF will be launched by a Titan IV-Centaur into a high earth orbit with an altitude of 100,000 km. It will operate at high efficiency, with tens of hours of uninterrupted coverage possible on a single area of sky. This initiative will unite two proven technologies to make a national observatory of unprecedented power. First, the technology for cooled telescopes has been demonstrated by two Explorer satellites, IRAS and the Cosmic Background Explorer (COBE). The thermal background for SIRTF will be a million times less than that for a terrestrial telescope. Second, SIRTF will take full advantage of the U.S.-led revolution in infrared detector arrays. IRAS had only 62 detectors; SIRTF will have over 100,000. Figure 4.1b shows SIRTF's expected sensitivity compared to the brightness expected for representative extragalactic objects. Figure 4.2 compares SIRTF with the European Infrared Space Observatory (ISO) mission, to be launched in 1994. Because of its larger aperture and its use of larger and more sensitive detector arrays, SIRTF will be thousands of times more capable than ISO. SIRTF will follow up on ISO's discoveries in addition to breaking new ground with its own deep surveys. All key technologies have been successfully demonstrated, either on the ground or with precursor space missions, during a decade of study by NASA. SIRTF could be initiated in 1994 and launched around 2000 to provide valuable overlap with the Hubble Space Telescope (HST) and the Advanced X-ray Astrophysics Facility (AXAF).

The committee's highest priority for ground-based astronomy is an 8-m-diameter telescope for the summit of Mauna Kea, Hawaii, optimized for low-background, diffraction-limited operation in the infrared between 2 and 10 μm but also useful in the optical regions of the spectrum. Mauna Kea is recognized as the best terrestrial site for an infrared telescope because of its low level of water vapor, the primary absorber and emitter of infrared radiation in the earth's atmosphere. Further, the remarkable stability of the atmosphere above this mountain results in minimal distortion of astronomical images. From this dry and stable site, an 8-m telescope would achieve diffraction-limited resolution of 0.1 arcsecond at 2 μm using modest "adaptive optics" techniques to correct for residual atmospheric distortion. The infrared-optimized 8-m telescope will gain its marked increases in sensitivity relative to the sensitivity of other large telescopes owing to the 0.1-arcsecond images possible in the infrared with adaptive optics, and to the low emissivity expected for a silver-coated monolithic mirror. The telescope would provide large amounts of data from new infrared detectors on a large telescope at high spatial resolution. Among the many projects the infrared-optimized 8-m telescope would carry out, two for which it would be particularly well suited are deep searches for and spectroscopic studies of primeval galaxies (Plate 4.2).

The proposed national infrared-optimized 8-m telescope will differ from

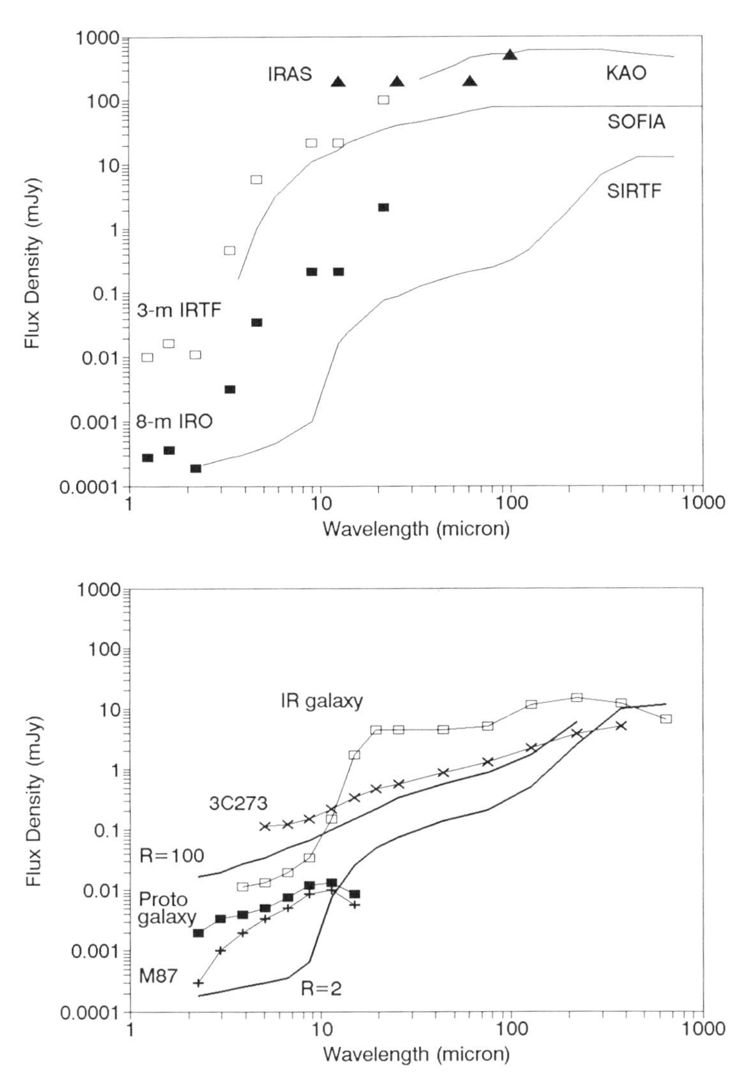

FIGURE 4.1a (top) Photometric sensitivity of three proposed facilities, SIRTF, SOFIA, and the infrared-optimized (IRO) 8-m telescope (solid squares) for broadband observations. For reference the figure also gives the sensitivity of the KAO, the 3-m IRTF (open squares), and the IRAS survey (solid triangles). Technical assumptions underlying these curves can be found in the report of the Infrared Astronomy Panel in the *Working Papers* (NRC, 1991).

FIGURE 4.1b (bottom) SIRTF's sensitivity (solid lines) for photometry [resolution (R) = 2] and spectroscopy [resolution (R) = 100] compared with the predicted brightnesses of representative extragalactic objects scaled to a common redshift of 5, when the universe was only one-sixth its present size. SIRTF could detect luminous galaxies, quasars, and protogalaxies in the early universe.

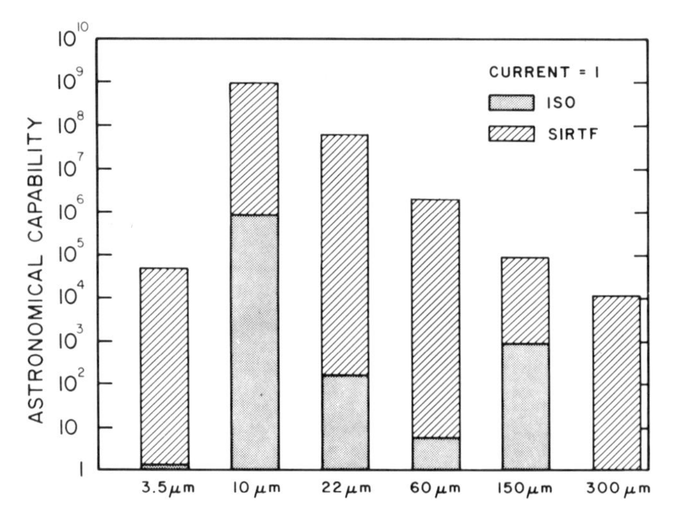

FIGURE 4.2 The power of SIRTF compared with existing telescopes and with the European ISO mission. "Astronomical capability" is defined as (facility lifetime) × (efficiency) × (number of detectors)/(sensitivity)^2 and is normalized at various wavelengths to the capabilities of NASA's 3-m telescope on Mauna Kea, IRAS, or the KAO. The figure shows how much more quickly SIRTF would be able to map or survey a region of sky to a particular flux limit than would other telescopes.

the private 10-m Keck telescope in three ways: (1) it will be accessible to all U.S. astronomers, whereas the Keck telescope will be available to only about 3 percent of the national astronomical community; (2) it will be the only large telescope in the world optimized for performance in the infrared; and (3) it will use adaptive optics to achieve the maximum possible spatial resolution in the near-infrared.

The third telescope needed to cover this factor-of-1,000 range in wavelengths is the Stratospheric Observatory for Far-Infrared Astronomy (SOFIA), a moderate-sized 2.5-m telescope. Mounted in a Boeing 747 aircraft, SOFIA will fly over 100 8-hour missions per year. At altitudes of 41,000 ft, above 99 percent of the water vapor in the earth's atmosphere, SOFIA will open the wavelength range from 30 to 350 μm to routine observation and make valuable contributions at still longer wavelengths. In particular, spectral observations at these wavelengths hold the key to understanding the physics in regions of high density and moderate temperature that characterize the primitive nebulae around newly formed stars, and the cores of infrared-luminous galaxies and quasars. SOFIA's capability for diffraction-limited imaging and high-resolution spectroscopy at wavelengths inaccessible from the ground will complement SIRTF's great sensitivity at infrared and submillimeter wavelengths. SOFIA is a joint project with Germany, which will supply the telescope system and support about 20 percent of the operations. NASA and the German space agency

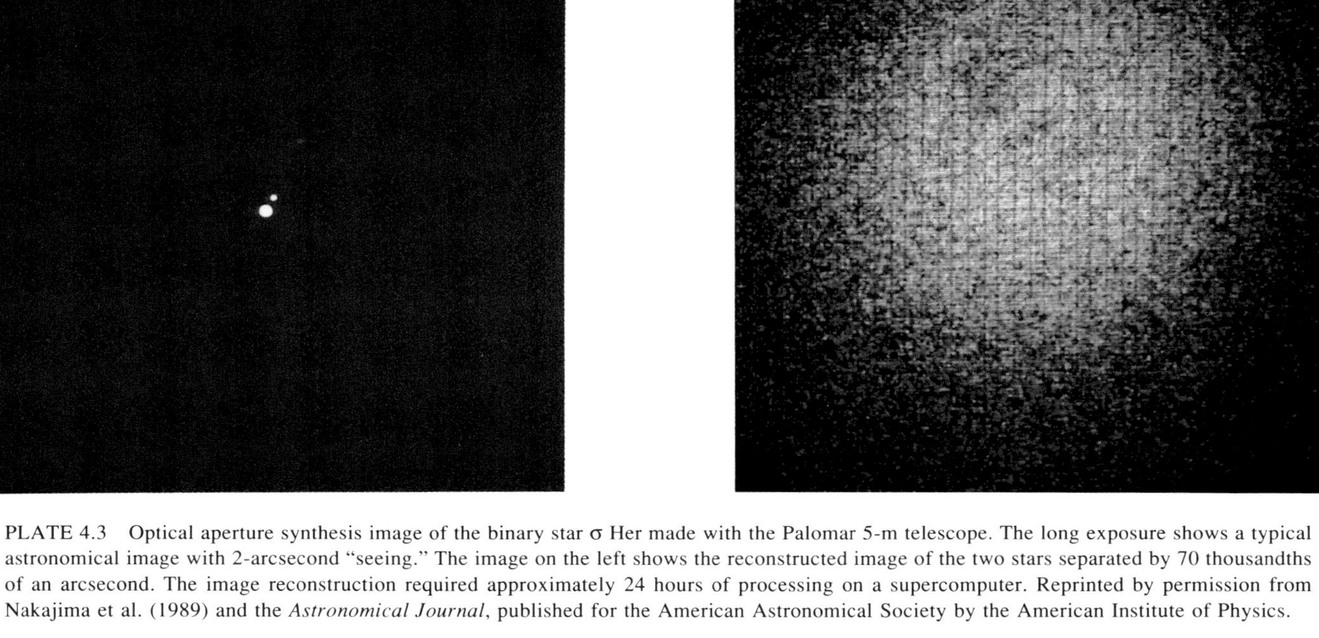

PLATE 4.3 Optical aperture synthesis image of the binary star σ Her made with the Palomar 5-m telescope. The long exposure shows a typical astronomical image with 2-arcsecond "seeing." The image on the left shows the reconstructed image of the two stars separated by 70 thousandths of an arcsecond. The image reconstruction required approximately 24 hours of processing on a supercomputer. Reprinted by permission from Nakajima et al. (1989) and the *Astronomical Journal*, published for the American Astronomical Society by the American Institute of Physics.

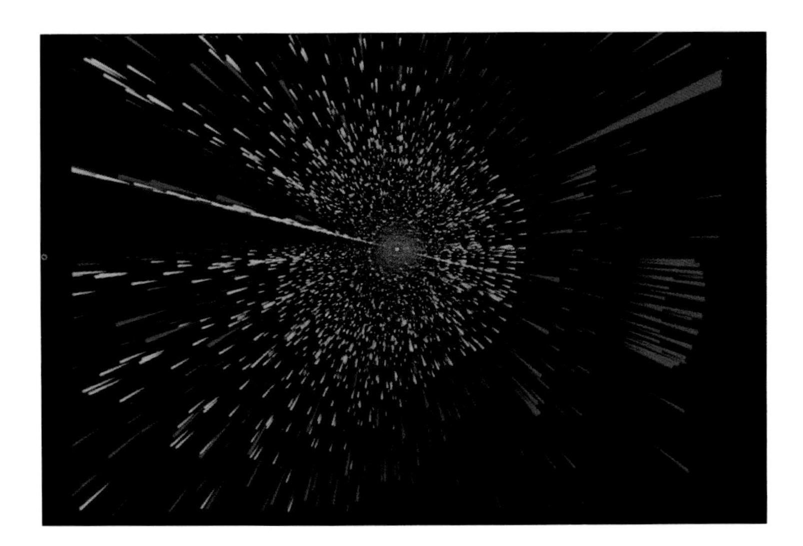

PLATE 5.1 What happens when two neutron stars collide head-on? Scientists programmed an NSF supercomputer with the known physics of neutron stars in an attempt to find out. The resultant smashup was spectacular. The streaks represent the trajectories of particles of stellar material. Courtesy of C. Evans, University of North Carolina, Chapel Hill, and C. Hoyer and R. Idaszak, National Center for Supercomputer Applications.

PLATE 5.2 The effect of a black hole is obtained from a numerical solution of Einstein's equations that describe the behavior of the gravitational field. The shape of the diagram measures the curvature of space due to the presence of the black hole. The color scale represents the speed at which a clock would measure time (red is slow, blue fast). Courtesy of L. Smarr, D. Hobill, D. Bernstein, D. Cox, and R. Idaszak, National Center for Supercomputer Applications.

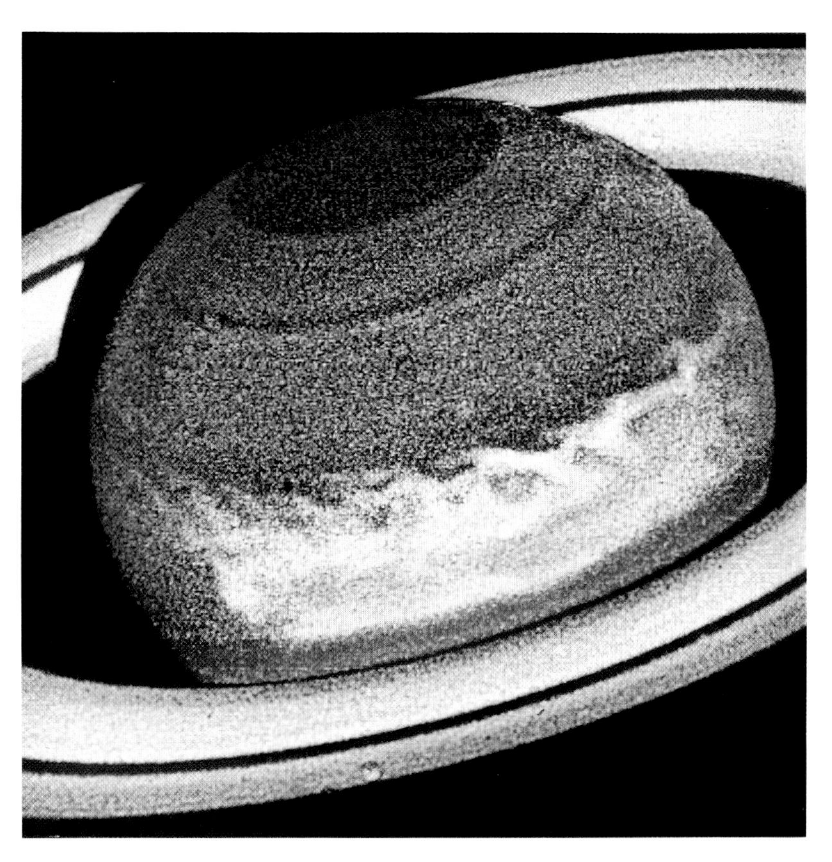

PLATE 8.1 This Hubble Space Telescope image of the planet Saturn was taken with the Wide-Field/Planetary Camera. The picture shows a storm system that grew from a small disturbance in the planet's Southern Hemisphere to an equatorial belt surrounding nearly the entire planet. The study of weather systems on other planets helps scientists understand weather patterns on the earth. Courtesy of NASA.

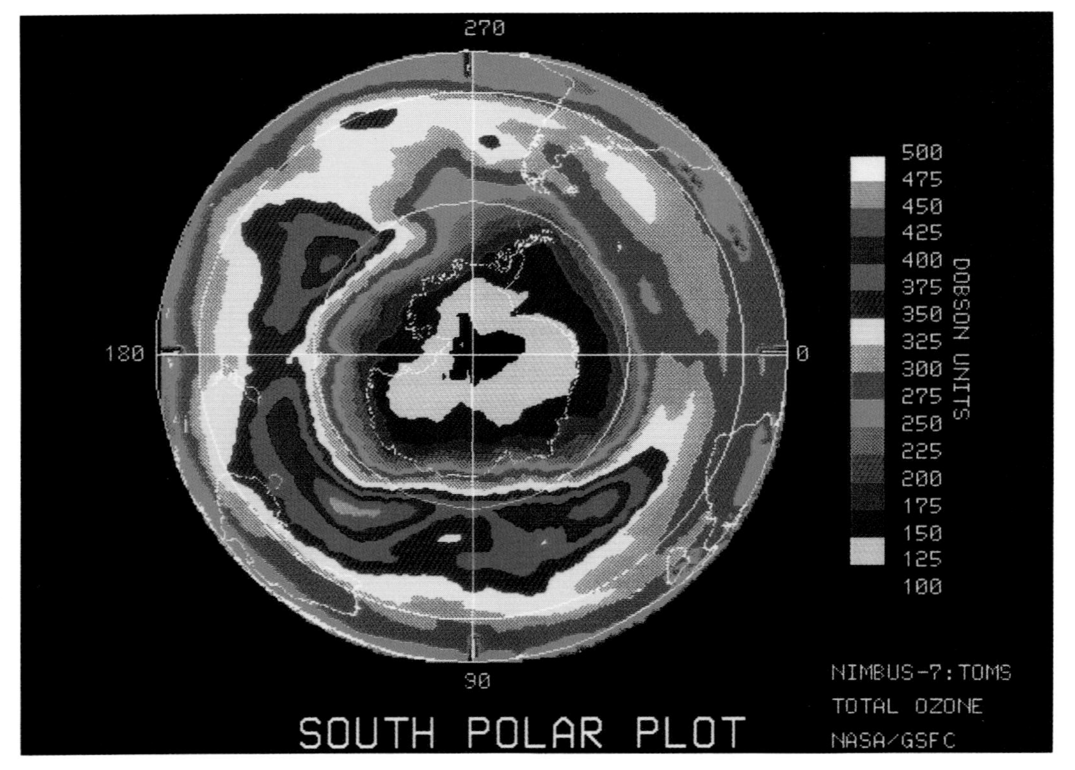

PLATE 8.2 An image (obtained October 5, 1987) of the terrestrial ozone concentration shows a hole over Antarctica. Ground-based and orbiting instruments and analytical techniques similar to those developed for the study of interstellar clouds and planetary atmospheres are used to study this important environmental problem. Image courtesy of NASA's Goddard Space Flight Center.

TABLE 4.3 Comparison of Different Recommended Infrared Facilities

Facility	Most Important Attributes
SIRTF	Unequaled sensitivity for imaging and moderate-resolution spectroscopy Broad wavelength coverage from 2 to 700 μm 7.5 ($\lambda/30$ μm) arcsecond imaging of faint sources at $\lambda > 30$ μm
Infrared-optimized 8-m telescope	0.7 ($\lambda/30$ μm) arcsecond imaging for $\lambda < 30$ μm High-resolution spectroscopy in atmospheric windows for $\lambda < 30$ μm Evolving instrumentation
SOFIA	High-resolution spectroscopy at $\lambda > 30$ μm 2.5 ($\lambda/30$ μm) arcsecond imaging at $\lambda > 30$ μm Training of instrumentalists

have successfully completed preliminary design studies for SOFIA, drawing heavily on the technical heritage of the KAO. SOFIA could begin observations in 1998.

Table 4.3 compares the most important attributes of the three recommended infrared facilities, SIRTF, the infrared-optimized 8-m telescope, and SOFIA. One can see from this comparison that the three instruments are mutually complementary.

SIRTF has the highest sensitivity for photometry, for imaging, and for low- to moderate-resolution spectroscopy ($\sim$100 km s^{-1}). Between 3 and 20 μm, SIRTF will be 10 to 40 times more sensitive than the infrared-optimized 8-m telescope. Despite advances in ground-based telescope design and detector technology, SIRTF will maintain fundamental advantages in sensitivity longward of 3 μm. SIRTF will also have the uninterrupted spectral coverage from 2 to 200 μm needed to detect important molecular and atomic spectral features.

The great strength of the infrared-optimized 8-m telescope compared with SIRTF will be its ability to operate at high spectral or spatial resolution, or both. The telescope, capable of subarcsecond resolution owing to its adaptive optics, will make maps with 100 times more spatial information than those made by the 0.9-m SIRTF telescope. Information on this angular scale will be critical for understanding the disks around young stars and the energy source of infrared-luminous galaxies. For spectroscopy shortward of 5 μm, and at resolving powers in excess of 100,000 from 2 to 20 μm, the infrared-optimized 8-m telescope will be more sensitive than a space-based one throughout the wavelength region accessible from the ground. The infrared-optimized 8-m telescope will make seminal contributions to problems requiring both high spatial and spectral resolution, such as probing the centers of dusty galaxies like our own to look for evidence of massive black holes.

Instruments on SOFIA will be able to operate with the spectral resolution of 1 km s^{-1} needed for dynamical studies of galactic sources in wavelengths inaccessible from the ground. SOFIA will make spectroscopic observations in the large variety of molecular and atomic transitions that characterize far-infrared and submillimeter wavelengths. For bright sources, SOFIA's maps will contain nine times the spatial information of maps made by SIRTF, although SIRTF will be much more sensitive for broadband observations. This information will be extremely useful in probing the physical conditions in protostars, planetary debris disks, and galaxies. Finally, SOFIA will provide excellent training for young experimentalists and a valuable opportunity to develop and test new instruments.

Other projects operating in the infrared and submillimeter region are discussed in the *Working Papers* (NRC, 1991) and include an Explorer mission to make a spectral survey of important classes of objects at submillimeter wavelengths; a ground-based survey of the entire sky at 1 to 2 μm with a threshold 50,000 times fainter than that of the only other survey, which is now 20 years old; allocation of funds to equip ground-based telescopes with revolutionary new infrared arrays; and a radio telescope using arrays of receivers to look for anisotropies in the cosmic background radiation that might provide clues to when and how galaxies formed.

HIGH SPATIAL RESOLUTION

Basic physical principles limit the smallest angle that a telescope can discern to a value approximately equal to the wavelength of the radiation observed divided by the characteristic size of the telescope. At visible wavelengths this limit is a few hundredths of an arcsecond for a 5-m telescope, although this limit has not been achieved until very recently, and then only under special circumstances, the limiting factor in practice being the degradation by turbulence in the earth's atmosphere. The highest quality of astronomical images at the best ground-based sites on the rarest, most stable nights is about 0.3 arcseconds. At radio wavelengths, the limiting resolution of the largest single antennas varies with wavelength from about five to a few hundred arcseconds. Radio astronomers have developed the technique of interferometry in which small telescopes spread over large distances are linked together to simulate a single telescope with an aperture equal to the largest separation between the component telescopes. The Very Large Array (VLA) consists of 27 telescopes separated by up to 35 km and is capable of resolving structures about 0.3 arcsecond in size. Very long baseline interferometry uses telescopes distributed over the entire earth to resolve objects smaller than a thousandth of an arcsecond.

Six of the programs recommended in Chapter 1 stress 10-fold or greater improvements in spatial resolution compared with that possible with existing facilities. At visible and infrared wavelengths new technologies may reduce the

FIGURE 4.3 An artist's conception of the proposed MMA shows the 40 8-m telescopes spread out in a ring 900 m in diameter. Inner and outer rings of 70 m and 3 km, respectively, are also visible. Courtesy of the National Radio Astronomy Observatory/Associated Universities, Inc.

effects of atmospheric distortion and lead to interferometers capable of resolving a few thousandths of an arcsecond. At radio wavelengths, interferometric techniques will make possible an array of telescopes capable of subarcsecond imaging at 1 mm (1,000 μm), bringing many classes of phenomena into clear view for the first time.

The Millimeter Array

The recommended Millimeter Array (MMA) will be a sensitive, high-resolution instrument providing high-fidelity images at wavelengths between 0.9 mm (900 μm) and 9 mm. Comprising 40 individual 8-m telescopes, it will have a resolution at 1 mm that ranges from 0.07 arcsecond for the largest (3-km) configuration of the telescopes, to 3 arcseconds for the compact (70-m) configuration (Figure 4.3). The technology for the MMA is well understood and draws on expertise developed with centimeter interferometers like the VLA, university millimeter interferometers, and commercial systems. Compared with any other millimeter telescope in the world, existing or planned, the MMA will have better angular resolution by more than a factor of 10, better sensitivity by a factor of 30, and better imaging speed by a factor of 100 or more. The MMA will provide observations in a wavelength regime and with spatial and spectral resolution that are highly complementary to the three infrared telescopes discussed above.

The MMA will aid the studies of galaxy formation by detecting the dust and gas emission from very young galaxies, early in the history of the universe. Surveys tracing emission from carbon monoxide molecules will lead to three-dimensional maps of the large-scale distribution of spiral galaxies out to cosmologically exciting distances. Images with high spatial and spectral resolution will unveil the kinematics of optically obscured galactic nuclei and quasars on spatial scales smaller than 300 light-years. Images of young stars taken with existing millimeter interferometers have already demonstrated the existence of enough material in orbit around younger versions of the sun to make 10 to 100 Jupiters. The MMA will measure the mass, temperature, and composition of such protoplanetary disks with greatly improved sensitivity and resolution.

Adaptive Optics

The recommended program in adaptive optics will give existing and future large telescopes the ability to remove atmospheric distortions, thereby increasing the resolution and sensitivity of astronomical measurements. "Adaptive optics" is different from "active optics." The latter refers to techniques, like those planned for the Keck 10-m optical telescope or the Green Bank 100-m radio telescope, to correct for minute- or hour-long drifts in mirror shape due to gravity, wind, and temperature drifts. "Adaptive optics," however, attempts to compensate for the rapid, hundredth-of-a-second effects of atmospheric turbulence. Of the two, adaptive optics is the more challenging, but also the more rewarding scientifically.

Turbulence in the atmosphere scrambles light waves in patches larger than a characteristic size, r_0, about 8 in. at visible wavelengths. Consequently, light reaching a telescope of diameter larger than this is so badly disordered that diffraction-limited imaging is normally impossible. European astronomers, as well as U.S. scientists, have developed techniques to monitor the wave-front errors in each r_0-sized patch of a telescope mirror, and to correct them with reference to a nearby standard star by warping the mirror appropriately. Corrections must be made within the "coherence time" of the atmosphere, τ_0, roughly every hundredth of a second at visible wavelengths. Complete phase corrections require a reference star brighter than a visual magnitude of about 9 located within an isoplanatic angle θ_0 of the object of interest. This angle is only a few arcseconds at visible wavelengths. Partial corrections can still improve angular resolution and might utilize stars as faint as 15 magnitude separated by larger distances.

The size parameter, r_0, the coherence time, τ_0, and the allowable distance between object and reference star all increase with wavelength. Thus adaptive optics will probably be applied first in the infrared. The number of r_0 patches across a telescope is much smaller in the infrared than in the visible, so that the number of correcting actuators is tens instead of hundreds. Corrections are needed less frequently, and fainter and more numerous reference stars can be used, since the coherence time is longer, up to half a second. Finally, more sky is available to search for reference stars because θ_0 can be as large as 100 arcseconds.

Ultimately, a solution to the problem of finding suitable reference stars at visible wavelengths may be to use lasers to generate "artificial stars" in the sky close to sources of interest. This technique has been partially developed by the Department of Defense but needs to be adapted to astronomy.

The committee proposes that adaptive optics techniques be developed and implemented on existing and planned large telescopes, such as NASA's existing Infrared Telescope Facility (IRTF) on Mauna Kea and on the new generation of 8- and 10-m telescopes. It should be possible to produce near-diffraction-limited performance at wavelengths as short as 1 μm. The proposed program would also support development and implementation of adaptive optics on existing solar telescopes, as pioneered at Sacramento Peak, and the development of adaptive optics for the proposed Large Earth-based Solar Telescope (LEST).

The scientific gains from applying adaptive optics will have an enormous impact on many branches of astronomy. The desired 0.1-arcsecond imaging resolution at 2 μm is adequate to resolve details on many planets and satellites in our own solar system, to examine nascent solar systems around young stars, to search for signs of black holes in the cores of energetic galaxies, and to look for supernovae in young galaxies at redshifts of 1 or greater. Furthermore, the reduction in background-noise contamination made possible with diffraction-limited optics could improve the limiting sensitivity of ground-based telescopes by up to a factor of 50.

Optical and Infrared Interferometers

In the 1920s A. Michelson and F. Pease showed that two or more telescopes linked together can achieve a spatial resolving power proportional to the distance between the apertures. Although such interferometers have been built with great success at radio wavelengths, relatively little progress has been made, until very recently, at optical or infrared wavelengths. In the past decade optical interferometers in the United States and Europe have produced useful scientific results, including wide-angle astrometry with thousandth-of-an-arcsecond precision, measurements of stellar diameters with a precision sufficient to constrain stellar atmosphere models, and the resolving of close binary stars. These projects have demonstrated that the key technologies are in hand to make interferometers using 1- to 2-m-diameter telescopes separated by baselines up to a kilometer, with corresponding resolution better than a thousandth of an arcsecond at a wavelength of 1 μm. Figure 4.4 and Plate 4.3 show two recent results from optical interferometry.

Continued support for existing small interferometers and the development

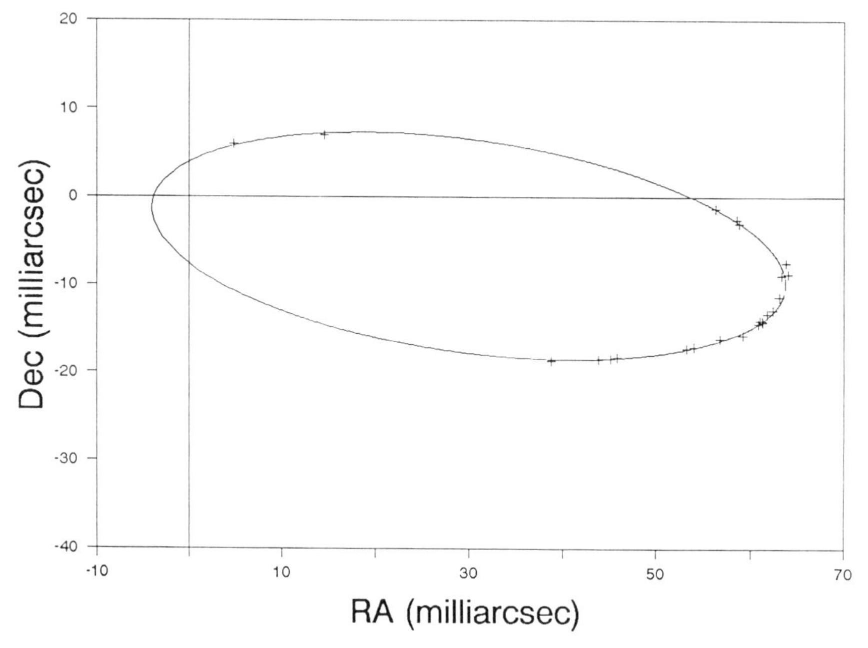

FIGURE 4.4 Apparent orbit of the binary star Beta Arietis observed with the Mark III interferometer operating on Mt. Wilson. The Mark III system has routinely demonstrated the ability to control the active optical elements to the hundredths-of-a-micron precision necessary for a scientifically productive instrument. Reprinted by permission from Pan et al. (1990) and the *Astrophysical Journal*.

of at least one larger one would yield immediate scientific return and would help solve technical problems. An array of five 2-m telescopes with baselines exceeding 100 m would provide infrared images of important astronomical objects with a resolution of a few thousandths of an arcsecond, and would lead the way for more ambitious programs in the next decade. The facilities operated in the 1990s would have a major impact on the study of individual stars, of newly forming stars, and of the compact cores of luminous galaxies that may harbor black holes. As discussed in Chapter 6, ground-based interferometers will provide tests of key technologies and concepts for future space interferometry and are an important part of a balanced program leading to major space facilities. Late in the 1990s the program would support planning and design of an advanced interferometer to be built in the next decade.

Astrometric Interferometry Mission

Despite the promise of interferometry carried out from the ground, the ultimate power of this technique will probably be fully realized only with a system operating in space, where no limits to the coherence size, angle, or time are imposed by the terrestrial atmosphere. NASA has been studying a number of possible space missions as a first step in interferometry from space. A concept that appears particularly appealing is to use interferometric techniques to achieve a 1,000-fold improvement in our ability to measure celestial positions. The mission requirement would be to measure positions of widely separated objects to a visual magnitude of 20 with a precision of 30 millionths of an arcsecond; a more challenging goal would be to measure positions with a precision of 3 millionths of an arcsecond. The Astrometric Interferometric Mission (AIM) would permit definitive searches for planets around stars as far away as 500 light-years through the wobbles of the parent star, trigonometric determination of distances throughout the galaxy, and the study of the mass distributions of nearby galaxies from stellar orbits. AIM would demonstrate the technology required for future space interferometry missions.

Large Earth-based Solar Telescope

The Large Earth-based Solar Telescope is a solar telescope with a 2.4-m aperture that would use adaptive optics to increase the spatial resolution of solar observations. It would be the premier terrestrial telescope for high-resolution solar observations at optical and near-infrared wavelengths. With LEST, it will be possible to investigate in unprecedented detail the interactions of magnetic fields and turbulent motions under way in the solar surface and overlying atmosphere, which are responsible for the hot solar corona, the solar wind, solar flares, and solar-terrestrial phenomena.

The program will be a cooperative international venture among nine countries. It will combine U.S. expertise in adaptive optics and instrument design with European contributions of an outstanding site in the Canary Islands and a major share in the costs of construction. The committee endorses a plan in which the United States would pay one-third of the construction and operation costs of LEST in return for a proportionate share of the observing time.

VLA Extension

The Very Large Array can currently produce images with better-than-arcsecond resolution. This will be complemented by the better-than-a-thousandth-of-an-arcsecond resolution of the Very Long Baseline Array (VLBA) when it is completed in 1992. There will still remain, however, a gap between the capabilities of the VLA and the VLBA that will restrict the performance of the combined instruments at intermediate resolutions between about 0.01 and 0.1 arcsecond. A plan to bridge this gap would be carried out in three phases: (1) supply the VLA with VLBA tape recorders, (2) build four new antennas to provide intermediate spacings between the VLA and VLBA, and (3) add fiber-optic links between the VLA and nearby VLBA antennas. Together these improvements will produce a genuine intercontinental telescope, one that combines the full angular resolution of the VLBA with the enormous sensitivity and dynamic range of the VLA. This will permit a wide range of new astrophysical applications, including, for example, rapid and high-angular-resolution measurements of solar flares, the imaging of both thermal and nonthermal emission from nearby stars, and astrometric observations with a precision better than a thousandth of an arcsecond. The high-resolution images of the internal structure of jets and lobes in radio galaxies and quasars obtained with this intercontinental telescope will provide a new tool for cosmological studies.

CONSTRUCTION OF LARGE TELESCOPES

Optics technology has progressed to the point that telescope makers are confident of being able to build successfully the first optical and infrared telescopes larger than the 5-m Hale telescope. Recent advances in the technologies for casting and polishing of fast mirrors, in the precise alignment and support of segmented and monolithic mirrors, and in simple altitude-azimuth mounting will enable the construction of 8- and 10-m telescopes and will make the construction of 4-m-class telescopes less expensive. The 8- and 10-m telescopes will bring 10-fold improvements in sensitivity and, when coupled with the adaptive optics techniques described above, will permit a remarkably sharp view of astronomical objects. Table 3.1 and Appendix B list the telescopes proposed or under construction for the 1990s. The Keck instrument nearing completion on Mauna Kea will employ a segmented primary mirror made up of 36 hexagonal mirrors locked together with an advanced servomechanism to operate as a single 10-m telescope (Figure 4.5). The 8-m Spectroscopic Survey Telescope will be even more highly segmented than the Keck, using 85 1-m-diameter spherical mirrors. The other planned 8-m telescopes will use lightweight monolithic mirrors fabricated at the Mirror Laboratory of the Steward Observatory at the University of Arizona (see Chapter 3) with new casting and polishing techniques. Much of the required technology has already been tested with two 3.5-m mirrors, and a 6.5-m telescope is now under construction for the Smithsonian Institution. If appropriate funding is secured, the Mirror Laboratory will make the 8-m mirrors for the public telescopes recommended in this report, as well as for some of the private ventures listed in Table 3.1.

A Southern 8-m Telescope

The second major ground-based telescope recommended by the committee is an 8-m telescope, optimized for operation at optical wavelengths, to be built in the Southern Hemisphere. The initial set of instruments would include

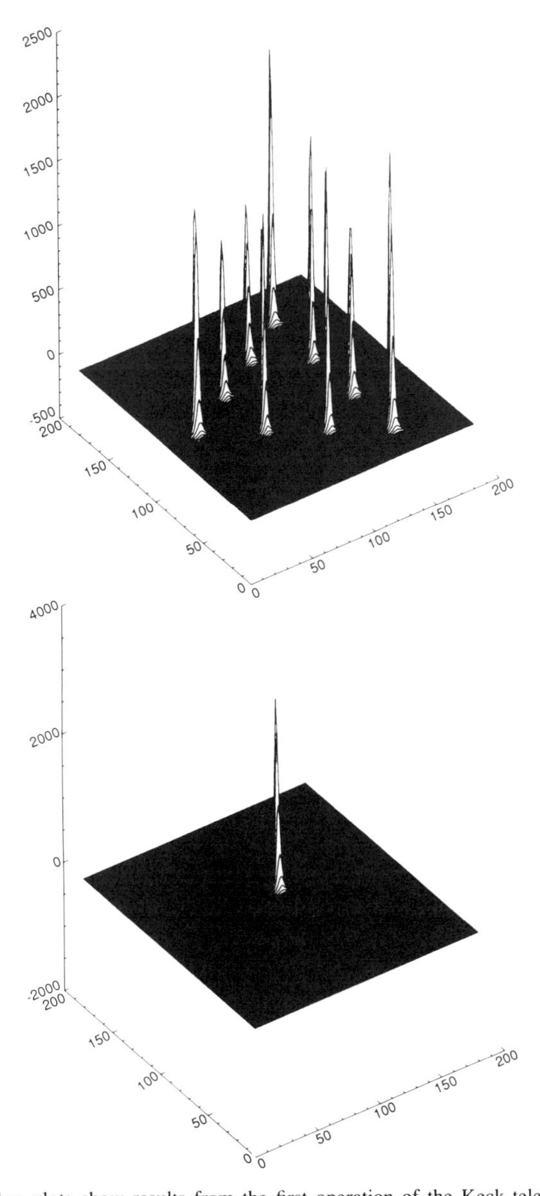

FIGURE 4.5 Two plots show results from the first operation of the Keck telescope in Hawaii. Above, individual images of a single star, separated by about 12 arcseconds, are formed by the nine mirrors that make up the partially completed telescope. Shown below is the result of the telescope's servosystem bringing the nine images together to form a single image; the resultant image is less than an arcsecond across. The two horizontal scales are in tenths of arcseconds; the vertical scale is in arbitrary intensity units. When completed in 1992, the Keck telescope will have 36 mirrors operating together as a single 10-m telescope. Courtesy of the California Association for Research in Astronomy.

multiobject and high-resolution optical spectrometers. Such a telescope would complement its Northern Hemisphere counterpart through its ability to carry out essential scientific programs for which an infrared-optimized telescope would be less suitable. The telescope should have a monolithic mirror of optical quality sufficient to take full advantage of developments in adaptive optics techniques.

The Southern Hemisphere 8-m telescope would provide U.S. astronomers with a vital window on objects that are uniquely or best observed from the Southern Hemisphere, such as the Magellanic Clouds, the galactic center, and some of the most prominent globular clusters and radio galaxies. Without national access to a Southern Hemisphere 8-m telescope, U.S. astronomers will be at a disadvantage in understanding new objects and phenomena discovered by NASA's orbiting observatories.

Construction and Support of 4-m Telescopes

At present, only a small fraction of the first-rate investigations proposed for the 4-m telescopes of the National Optical Astronomy Observatories can be granted time, and often the time available is so minimal as to preclude ideas and programs with high risk but potentially great return, or the assembly of databases adequate to ensure proper interpretation. This pressure will increase as discoveries made with 8-m telescopes and the space observatories place insuperable demands on existing facilities for supporting observations.

Examples of the range of important scientific programs requiring extensive time on 4-m-class telescopes include characterizing, through imaging and spectroscopy, the physical properties of sources discovered at nonvisible wavelengths; determining the interior structures of stars through long-term programs of spectroscopic monitoring of stellar oscillations; searching for planetary systems and subsolar mass objects by means of long-term radial velocity studies of large samples of stars; carrying out statistically complete spectroscopic and photometric studies of supernovae in galaxies and in active galactic nuclei; determining the mix of stellar populations in galaxies of a wide variety of ages and morphologies; mapping the large-scale structure of the universe out to a distance of 1 billion light-years (corresponding to redshifts of $z \sim 0.1$) by determining the redshifts of a million galaxies; and mapping the large-scale structure of the universe far beyond a billion light-years (corresponding to redshifts of $z \sim 1$) by determining redshifts of galaxies in carefully selected areas.

Fortunately, the technological advances that enable the construction of 8-m-class telescopes have also greatly reduced the expected size, weight, and cost of 4-m telescopes, while enhancing their image quality and operational efficiency. The superb image quality obtained with the New Technology Telescope of the European Southern Observatory attests to the potential of these new facilities. As a result of these advances, powerful telescopes can be built by individual universities or small consortia of institutions. The committee points out that strong university support of these facilities is essential to reduce costs, to involve students in important scientific programs, and to develop novel instrumentation.

THE INFORMATION EXPLOSION

Astronomers of the 1990s will collect an enormous volume of data with the widespread use of large-format electronic detectors and arrays of antennas. Many existing or planned telescopes will be equipped with instruments capable of producing tens of gigabytes of data per day in the form of sky surveys, multiobject spectra, three-dimensional spectroscopic images, and multidimensional polarization maps. Chapter 5 discusses these areas in more detail and makes specific recommendations regarding the computers, software, and procedures needed to obtain, process, and interpret these data, and about the necessity of electronic archives to make selected datasets generally available.

OTHER INITIATIVES

Progress in a field as diverse as astronomy cannot be completely summarized by just a few technological or scientific themes. The remaining recommendations of Chapter 1, not discussed in previous sections, will produce important scientific results in various areas of astronomy.

Dedicated Spacecraft for FUSE

A decade ago the Field Committee listed as its highest-priority moderate program a far-ultraviolet spectrometer in space. In 1989 NASA selected the Far Ultraviolet Spectroscopy Explorer (FUSE) mission to enter the Explorer queue for launch sometime around 1999. The committee strongly endorses the scientific importance of FUSE and recommends as its first choice in the moderate space category that this timetable for FUSE be ensured, and possibly accelerated, by the purchase of a dedicated spacecraft. As currently conceived, FUSE would be the third payload to use the single, reusable Explorer platform. FUSE would be carried aloft by the Space Shuttle and exchanged by Shuttle astronauts for the X-ray Timing Explorer (XTE) on the orbiting Explorer platform. A few years prior to this, the same procedure would have been used to replace the Extreme Ultraviolet Explorer (EUVE) with XTE. As discussed in Chapters 1 and 7, the committee believes strongly that coupling Explorer missions to the manned space program will lead to unnecessary delays and expense compared with launching such satellites on Delta rockets. Additional advantages of a dedicated Delta launch would be an increase in the on-orbit lifetime of FUSE and an optimized orbit that would improve its operational efficiency.

Acceleration of the Explorer Program

The goal of an accelerated Delta-class Explorer program is to fly a total of six astrophysics Explorers during the 1990s, two or three more than are currently planned. This recommended flight rate is considered essential for a revitalized program of moderate missions, as discussed in Chapter 7. Although specific missions should be selected by the normal peer review process, the committee believes that three areas of space astronomy are particularly primed for Delta-class experiments: gamma-ray spectroscopy of galactic and extragalactic sources, a complete submillimeter line survey of important astronomical objects, and an x-ray telescope capable of making images with 60-arcsecond resolution in the energy range 10 to 250 keV. These and other possible Explorers are discussed in the *Working Papers* (NRC, 1991). A related recommendation is the acceleration of the Small Explorer (SMEX) program from its currently planned two or three astronomy missions in the 1990s to a total of five missions in the decade.

Fly's Eye Telescope

Cosmic-ray protons with energies greater than 10^{19} eV are not confined by the galactic magnetic field, so that their observation can reveal their point of origin, either galactic or extragalactic. The existing Fly's Eye telescope in Utah (see Plate 2.9) has detected some 200 fluorescent trails of highly energetic cosmic rays (10^{19} to 10^{20} eV) moving through the atmosphere. The direction, energy, and longitudinal development of the airshower can be measured. The present data suggest an isotropic distribution of particles with a flattening of the spectrum at energies above 10^{19} and a possible cutoff at energies above 10^{20}. The longitudinal development of the airshower suggests that these particles are protons. These particles may come from outside of the galaxy, but few mechanisms are known that can accelerate particles to these energies, and no mechanisms are known that can fill the universe with such energetic particles. However, the statistics on which these conclusions are based are sparse. A new Fly's Eye telescope would be 10 times more sensitive and would detect many more events than the existing instrument. The statistics of more than 2,000 events in a few years would lead to a better determination of the energy spectrum and the isotropy of these energetic cosmic rays. The improved spectrum would help determine whether the cutoff at energies of 10^{20} eV, expected from pion-producing interactions of protons with the 2.7 K cosmic background radiation (the Greisen effect), is real. The improved spatial resolution would be used to make more detailed studies of the longitudinal development of the airshowers and thereby infer the composition of the particles. This modestly priced facility will explore a new domain in cosmic-ray physics and could yield fundamental new insights.

5
Astronomy and the Computer Revolution

INTRODUCTION

Computer technologies have been central to advances in astronomy and astrophysics for the last 40 years and will play an even more important role in analyzing more complex phenomena in the next decade. In the early 1950s, roughly half the cycles of John von Neumann's pioneering MANIAC computer were devoted to the first stellar evolution codes. In the 1960s, advanced computers allowed the first detailed models of supernova explosions. In the 1970s, the Einstein Observatory x-ray telescope and the Very Large Array (VLA) of radio telescopes created images using computers as intermediaries between the telescope and the observer. In the 1980s, microcomputers came into wide use for the control of data acquisition at telescopes, and theoretical simulations were extended to a wide variety of complex astrophysical phenomena. In the 1990s, astronomers will apply powerful new computer technologies to obtain, process, and interpret the data from ground- and space-based observatories. The field of astronomy will, by virtue of its dependence on large quantities of data and its past experience and future goals, be the leader in important aspects of a national program in high-performance computing.

In this chapter, the committee discusses exciting developments in astronomy that can occur as a result of enhancements in computing strategies, techniques, or power. Since computing has a central role in astronomical research, the committee makes recommendations regarding archiving, workstations and hierarchical computing, networks, and community code development. The costs for computer support are not given explicitly in this chapter because the costs occur in various programs that are discussed in Chapters 1, 3, and 4. For example, workstations are included in the grants program, supercomputing support is provided directly to the supercomputing centers, and computers to perform the first stage of data reduction are included in the costs of individual instrumental initiatives.

A HIERARCHY OF COMPUTING POWER

A hierarchical network of ever more powerful machines will provide great computing power to the individual researcher in the next decade. Beginning in the 1980s, personal computers and workstations gave many scientists control over their own computing and observing environments; supercomputers became generally available through the creation of the NSF national supercomputer centers; and international research networks allowed researchers to communicate electronically with their colleagues or with supercomputer centers, national laboratories, observatories, and data archives. These trends will accelerate and become more tightly integrated in the next decade.

The 1990s will bring major advances at all levels in this evolving hierarchy. New processor technologies will put affordable, powerful computing in every observatory and on every astronomer's desk. These machines, which are nearly as powerful as today's supercomputers for many tasks, will make possible the acquisition and processing of large datasets, as well as the forging of synergistic links between data analysis, theoretical computations, and visualization tools. On a slightly larger scale, departmental mini-supercomputers will offer performance enhancements over desktop machines and provide sharing of expensive peripherals. Optical recorders capable of storing more than 100 terabytes each will hold archives of important datasets from ground- and space-based telescopes. Electronic networks will facilitate scientific collaboration and provide access to the national archives of observational and laboratory data. Despite the advances in local computing, there will still be a place for large central computers: multimillion dollar supercomputers will be bigger and faster versions of whatever is sitting on one's desk. These machines, with their huge memories, extensive disk capacity, and extremely fast processing and input and output rates, will be crucial to a minority of computer users with the most demanding programs.

DATA ACQUISITION AND PROCESSING

Computers have become essential to the acquisition of astronomical data, offering enhancements in performance comparable in some cases to improvements due to new telescope and detector designs. Software engineering has become as important to the success of a new instrument as are mechanical, electrical, or optical engineering. High-performance computing has become necessary to make full use of many observations. For example, powerful algorithms have greatly enhanced imaging telescopes that operate in the radio and in the optical domains (Figure 5.1; see also Plate 4.3). The VLA that astronomers use today is much more powerful and flexible than the one that was originally designed, without any major modification of the telescope itself, because of more powerful processing at the VLA and at national supercomputer centers.

The volume of data produced by some astronomical instruments is already large and is growing rapidly. Data rates of 10 gigabytes per day are common, and 100 gigabytes per day may soon be exceeded. Detector arrays will soon produce two-dimensional images with up to 2,048 × 2,048 pixels (elements), and some instruments will add spectral, temporal, or polarization channels to generate even larger datasets. The data flow from the VLA can exceed 72 gigabytes per day. The current VLA computers cannot handle this maximum data flow, so that subsets of the data must be selected for transmission and analysis. This under-utilization could become even more dramatic for the next generation of optical and infrared telescopes that will use arrays consisting of millions of individual detectors. Computers capable of processing 30 gigabytes per night will be as essential as improved detectors for the optical and infrared sky surveys planned for the 1990s. Without adequate computational and data storage capabilities, astronomers will not be able to push current or planned ground-based telescopes, which cost tens of millions of dollars each, to their limits.

DATA REDUCTION AND ANALYSIS

Intensive data processing is often required to convert observations to understanding. The type of data processing required varies widely depending on the telescope and the purpose of the observations. Large user facilities with stable instrumentation and operating conditions like the VLA lend themselves to processing in which the typical user has little involvement until the final analysis stages. Similarly, specialized survey instruments like the Infrared Astronomical Satellite (IRAS) need production software to process large volumes of raw data into scientifically useful catalogs and images. Software development for these instruments is often best handled by professional software engineers working in close conjunction with astronomers who understand the technical problems and scientific goals and who are familiar with details of the instruments.

The problem of making general-purpose data-reduction software is a difficult one. The National Optical Astronomy Observatories developed the Image Reduction and Analysis Facility (IRAF) package as a community program for reducing, calibrating, and analyzing images or two-dimensional spectra from optical and infrared telescopes. Similarly, the National Radio Astronomy Observatory created the Astronomical Image Processing System (AIPS) for

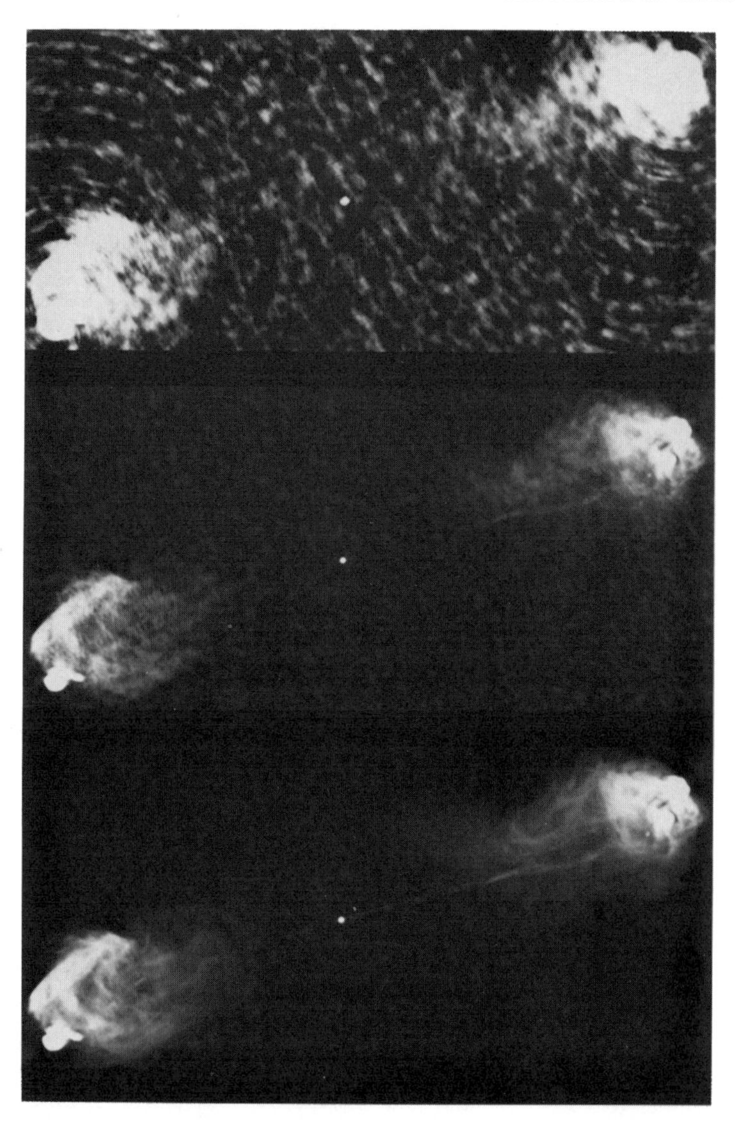

FIGURE 5.1 Three images of the radio source Cygnus A, a luminous, nearby radio galaxy, that represent different stages in processing of data from the VLA. At the top is the initial map produced with the VLA after a minimum of computer processing. Defects in the image due to residual effects of the atmosphere and other calibration errors manifest themselves as mottled structures all over the image. Additional processing steps in the middle and bottom frames "clean" the image in a mathematically well-defined way and enhance the astronomer's ability to discern weak structures in the presence of strong sources by a factor of 100 or more. The largest images may require a supercomputer for complete "cleaning." Courtesy of the National Radio Astronomy Observatory/Associated Universities, Inc.

processing data from radio interferometers. Many astronomers have adopted IRAF or AIPS in preference to the daunting task of developing their own codes. But researchers who want to make novel or demanding uses of state-of-the-art instruments find general-purpose software stifling; they may need a special calibration algorithm for their instrument that is not in the package. These astronomers find it difficult to integrate innovative data-reduction techniques into a large, centrally maintained package.

No perfect solution to this problem is yet in hand. To make effective use of the enormous advances in computer hardware and astronomical instrumentation expected in the next decade, astronomers will need an accompanying development of scientific software. Standard packages with modern interfaces are needed for users of optical and infrared telescopes whose needs are relatively standard. Programs consisting of 100,000 lines of code often require tens of person-years to develop, followed by a comparable and continuing effort to modernize, maintain, and document. Yet such packages are often difficult to modify for researchers interested in extracting the maximum possible information from their data. New architectures are needed that will provide both an open framework, within which an innovative user can develop and subsequently share new techniques, and a powerful set of fundamental tools for the general user. The development of such programs represents a major challenge to astronomers and to computer professionals. It is vital for both NASA and NSF to augment their efforts in the development, maintenance, and augmentation of community software.

ARCHIVING

There are compelling scientific reasons for archiving selected astronomical data from ground- and space-based telescopes. First, astronomical processes occur on time scales that are long compared to the lifetimes of individual researchers. The history of astronomy includes many instances when archival material dating back decades, and sometimes centuries, has proven essential in solving modern problems. Astronomers have an obligation to preserve contemporary data in an intelligible form for future generations of astronomers. Second, it is widely recognized that the definitive interpretations of much space astrophysics data are not found in the first papers, but rather in more extensive archival studies. Third, the large two-dimensional detectors of the 1990s will produce images of the sky taken at different wavelengths and at different times. Archival researchers can reanalyze these rich datasets to answer new questions and to make serendipitous discoveries.

The volume of data that will be gathered in the 1990s amounts to many terabytes per year. NASA's astronomy missions alone will generate about 10 terabytes of data annually. NASA is setting up the Astrophysics Data System to ensure wide and rapid access to data obtained from space observations;

the system will use high-speed networks and a centrally maintained directory. NASA has a scientifically productive program for archival research, primarily to analyze data from IRAS, the Einstein Observatory, and the International Ultraviolet Explorer (IUE).

The decentralization of ground-based observatories and the mixture of private, state, and federal funding of research has made it difficult to establish archives of ground-based astronomical data. However, the recent development of high-speed networks and data format standards can simplify the archiving process. Data obtained electronically at ground-based observatories can now, in principle, be archived and made available to remote users over computer networks.

The U.S. observatories are lagging behind major foreign observatories, such as the Anglo-Australian Telescope, the European Southern Observatory, and the La Palma Observatory, in developing archiving programs. The obstacles in setting up archives in the United States are financial and cultural: few people are willing to expend scarce personal and fiscal resources on archiving their data, and most ground-based astronomical data obtained outside the national observatories are treated as the private property of the observer, with no imperative to turn the data over to the community. Overcoming these barriers will depend on developing suitable incentives for individual scientists to archive their data, on protecting the rights of the original observers, and on making the process as effortless as possible.

As a long-term goal, digital data from ground- and space-based telescopes should be archived in a scientifically useful form; the archives would include all the raw data, calibration data, and information necessary to remove instrumental signatures from the data. These data would be available to the community after an appropriate proprietary period, typically one to two years after the completion of an observing program. On-line archives of major observational datasets, catalogs, and processed data from the astronomical literature would improve productivity and enhance the return from both ground- and space-based science programs. A national archiving program would allow researchers and students from smaller colleges and universities to work with data from the best instruments in a way that is now impossible.

The cost of the relevant technologies is being reduced to the point that it is becoming realistic to store selected and appropriate data from major ground-based telescopes. The recent development of computer networks and the implementation of NASA's Astrophysics Data System provide both the means and the model for widespread access to archives of ground-based observations. New archives should be compatible with, and a part of, the directory service of NASA's Astrophysics Data System. The sums that NASA has invested in archiving are small in comparison to the overall cost of its missions, but large by the standards of ground-based astronomy. The NSF can take advantage of NASA's investment in this area to help set up an appropriate archiving program for ground-based astronomy. The committee notes that the NSF Subcommittee on Ground-based Optical and Infrared Astronomy, of the Advisory Committee for Astronomical Sciences, has highlighted digital archiving as a key recommendation.

COMPUTERS AND THEORETICAL ASTROPHYSICS

The light from many astronomical objects is produced by violent, complicated, and quickly evolving phenomena. Sophisticated simulations are often needed to understand, for example, shock waves around protostars and jets from the cores of active galaxies, but these simulations are frequently oversimplified due to a lack of sufficient computer resources. In the coming decade, more realistic simulations will play an essential role in understanding the underlying physics of these phenomena. The 1990s will be the decade in which a number of long-standing astrophysical problems will be solved, and computers will play an important role in these solutions.

Astrophysics depends on theory and modeling to a greater degree than most other physical sciences, because astronomers can observe only remotely. Moreover, the observed phenomena, the photons and fast particles that escape from astrophysical objects, are typically the result of complicated interactions among nonlinear processes. It is often necessary to construct elaborate models to achieve a satisfactory interpretation of the observations. More powerful computers and computer programs that incorporate realistic physics will greatly increase the ability of astrophysicists to extract physical insight from their observational data.

The committee estimates that 10 percent of practicing astronomers are engaged in theoretical simulation of astrophysical phenomena. Some of this computational astrophysics uses local workstations and mini-supercomputers. Roughly 10 percent of the time devoted to scientific computing at the NSF Supercomputer Centers is used by astrophysicists who are using supercomputers to solve problems that push the system to the limits of today's software and hardware capabilities (Plates 5.1 and 5.2).

The observational community has made concerted efforts at developing community software; commendable examples include NSF's development at its national observatories of AIPS, IRAF, and the Flexible Image Transport System (FITS) image format. Comparable efforts should be made in the area of theoretical astrophysics. A wide range of theoretical techniques must be developed to meet different problems in astrophysics, but it is difficult to predict a priori which techniques will be most useful. Large programs for stellar evolution, radiative transport, magnetohydrodynamic simulations, and characterization of material properties such as equations of state and grain opacities can take tens of person-years to develop. Community modeling tools would result in more efficient use of the individual astronomer's time and computer resources.

RECOMMENDATIONS

Archiving

- **The committee recommends that archives of selected digital data from ground- and space-based observations should be established and made available over a high-speed national network.**

The committee commends NASA's important steps in this direction for space-based data. Digital archiving is becoming increasingly important in sciences other than astronomy. Early funding for archiving could make astronomy a test bed for other disciplines. The NSF astronomy program should seek joint funding with other NSF divisions or from NSF management for a pilot archiving program but should proceed with such a program even in the absence of shared funding.

- **The committee recommends initiating promptly a technical study on the archiving of appropriate digital data from ground-based telescopes.**

The study might identify the requirements for, and a preliminary design of, a national digital distributed astronomical archive. Topics that could be considered include what, initially, would constitute appropriate datasets for archiving; proposals to ensure the broadest acceptance of archiving plans; the identification of costs and funding sources for archiving; community comments on the archive designs; integrating ground-based archives with NASA's Astrophysical Data System and Planetary Data System; interagency and international collaboration; and establishment of a national archive center. It is important that the astronomical community determine the relative priorities between archiving particular datasets and supporting new observations.

- **The committee endorses NASA's goal of archiving data from all space-based telescopes and recommends that a complementary and compatible archive of selected data from ground-based telescopes be initiated by the mid-1990s.**

The archiving initiative for ground- and space-based astronomy should have the following long-term goals: all new major observatories, both public and private, should incorporate an appropriate level of archiving in their design and in their standard operations; most public and private observatory surveys should be archived; and observers should obtain their raw data, whenever possible, in the archive format.

Workstations and Hierarchical Computing

- **The committee recommends the purchase of individual workstations and departmental mini-supercomputers to provide a distributed network for astronomical computing.**

NASA has, on the one hand, pioneered a cost-effective policy of funding workstations for research. On the other hand, computers have in some cases been eliminated from the budgets of approved individual NSF (and NASA) proposals. The committee believes that requests for workstations should be encouraged within the existing individual grants program; the workstations could include "basic" or "high-performance" machines according to the legitimate needs of the investigator.

The committee urges that the NSF promote mid-range, local computing by funding computers to be used jointly by small groups of investigators. A modest annual budget could provide 50 or more machines for theoretical and observational groups, the largest of which could be funded on a cost-sharing basis with the individual institutions. This investment would bring immense improvements to the people who rely on computers for their work. The committee also urges NASA to allocate an appropriate fraction of the Mission Operations and Data Analysis funds associated with the space missions of the 1990s for purchases of mini-supercomputers by those university departments involved in space research. The committee bases this suggestion on the assumption that the NSF, NASA, and DOE supercomputer centers, which must be periodically upgraded to remain at the forefront of technology, will continue to provide this scarce resource to the astronomical community.

Networks

- **The committee recommends that the NSF's Astronomy Division encourage the development of high-speed national networks by funding, on a continuing basis, links between the national networks and widely used observatories, interested astronomy departments, and other research groups.**

Community Code Development

- **The committee recommends that NASA increase its role in furthering the development of community software and software standards.**

6

Astronomy from the Moon

ASTRONOMY AND THE SPACE EXPLORATION INITIATIVE

According to current plans for the manned space program, humanity's return to the moon is not expected to take place until sometime in the first decade of the 21st century; substantial scientific facilities will not be established until even further in the future. Therefore this chapter does not recommend specific projects. Rather, discussion focuses on the moon as a site for astronomical telescopes and the science that may be best done with lunar telescopes. The committee's principal conclusion is that the moon is potentially an excellent site for some astronomical observations. The committee believes that a lunar astronomy program should complement the earth-orbiting satellite program, that both technology and science should proceed in a step-by-step fashion, and that NASA should devote an appropriate fraction of the funding for its Space Exploration Initiative to scientific endeavors, including astronomy. The committee outlines an evolutionary program that will develop necessary technologies and increase the scientific return from a lunar program.

A number of conferences have been held under NASA's auspices on the topic of lunar observatories (Burns and Mendell, 1988; Mumma and Smith, 1990). The reader is referred to these conference proceedings for many stimulating ideas. In the discussion that follows, the term "telescopes" is used in the general sense to include interferometers and astronomical instruments at all wavelengths, and equipment for the detection of cosmic rays.

THE MOON AS AN OBSERVATORY SITE

Physical Characteristics

The moon is a slowly rotating spacecraft, 3,476 km in diameter, that always presents the same face to the earth. The moon lacks a significant atmosphere but has a rocky surface covered with dust. Its surface gravity is about one-sixth that of the earth. Table 6.1 describes a typical infrastructure and compares it to other remote observing sites. Table 6.2 lists some of the advantages and disadvantages of the moon as an observatory.

The moon has most of the advantages of any observatory in space. The absence of a lunar atmosphere and ionosphere permits observations over the entire electromagnetic spectrum with a resolution that is limited only by the characteristic size of the telescope. The lunar environment also lends itself to the construction of large, precise structures. The low lunar gravity and the absence of wind make possible telescope mirrors and support structures lighter than those constructed on the earth. The lunar night provides thermal stability, important for maintaining the precise alignment of a large telescope or the separations and orientations of an array of smaller ones. During the lunar night, telescopes can attain the low temperatures, less than 70 K, needed to improve infrared sensitivity. A major advantage of the moon compared with orbital observatory sites is the large, rigid lunar surface on which could be built arrays of telescopes extending over many kilometers to form interferometers.

The disadvantages of the moon compared with a remote site on the earth or in earth orbit include the limited mass that can be sent to the moon, the stringent requirements imposed on the design of instruments that must survive the rigors of space travel, and the need for assembly and operation of complex equipment with only a few workers. A rocket that can send 1,000 kg to low earth orbit, or 400 kg to high earth orbit, can send only 290 kg to the moon. Although lunar gravity is weaker than the earth's, supporting a few tons of telescope is a difficult task not faced by the designer of an orbiting telescope. Cosmic rays and the solar wind impinge directly on the lunar surface, unmoderated by a magnetosphere. Contamination of optical and mechanical components by lunar dust is a potential problem.

Detailed study will be required to determine whether, for any particular instrument, operation from high earth orbit offers advantages relative to operation on the moon. In the very distant future, mining or manufacturing operations on the moon might create an infrastructure that could make the moon an attractive site for many astronomical facilities.

A Human Presence

The presence of astronauts offers both advantages and disadvantages for astronomy. Astronauts are able to install and repair astronomical facilities, albeit on a restricted work schedule and with dexterity limited by spacesuits. The direct

TABLE 6.1 Infrastructure at Remote Observing Sites

Operation Phase	Date	Science Mass (kg yr^{-1})	Power Available (megawatt)	Workers on Site
Antarctica	1990	5×10^5	0.35	100 (summer) 20 (winter)
Space Station[a]	2000	1×10^5	0.1	8
High earth orbit[b]	2000	4×10^4	0.1	10 (assembly) 0 (operations)
Lunar emplacement[c]	2004	2×10^3	0.1	4
Lunar consolidation[c]	2010	7×10^3	0.5	8
Lunar utilization[c]	2015	3×10^3	1	12

[a]Assumes two science payloads per year with a heavy lift vehicle.
[b]Assumes two science payloads per year using a heavy lift vehicle and assembly in low earth orbit, followed by boost to high earth orbit.
[c]*Report of the 90-Day Study* (NASA, 1989).

TABLE 6.2 Environmental Attributes of the Moon

Lunar Feature	Advantages	Disadvantages
No atmosphere	Access to all wavelengths No atmospheric distortion of images No wind loading of telescopes	No protection from cosmic rays No moderation of thermal effects
No ionosphere	No long-wave radio cutoff	Line-of-sight transmission
Size	Large, disconnected structures can be built Momentum from pointed telescopes is absorbed Seismically quiet compared to the earth	
Solid surface	Radiation and thermal shielding Raw construction materials	Possible dust contamination
Lunar gravity	Lightweight structures possible	Telescopes require support
Slow, synchronous rotation	Two weeks of thermal stability Long integration times Far side isolated from terrestrial interference	300 K diurnal temperature change Very slow aperture synthesis No solar power at night
Distance from the earth	Long baseline for radio interferometry	Expensive transportation
Human presence	Construction, operation, repair, refurbishment	Expense, safety requirements, environmental degradation

involvement of people in astronomical facilities mandates safety requirements that have, in the past, proven to be expensive. Several experiments done during Apollo missions showed that contamination problems due to the presence of humans may be acute on the moon. Previous experience suggests that scientific facilities should be designed so that humans are called on to provide only limited, but critical, services.

SCIENCE FROM A LUNAR OBSERVATORY

Some measurements for which the moon may offer significant advantages compared with terrestrial or earth-orbiting instruments are discussed in this section. The concepts described below are illustrative; other promising possibilities may be developed as plans for a lunar base mature.

Observations with Single Telescopes

The early operation of a modest-sized (1-m-class) telescope would provide vital information concerning the design of future, more complicated, lunar telescopes, while providing unique scientific information. A small pointed telescope could observe individual objects or carry out wide-angle surveys, perhaps in the ultraviolet or infrared. A transit telescope with only a few moving parts could produce an imaging survey from the ultraviolet to the infrared over a large area of the sky using the slow rotation of the moon for scanning. The telescope would provide a high-resolution map of the universe at faint magnitudes.

A large-diameter telescope (16-m class) operating at infrared, optical, and ultraviolet wavelengths could have enormous scientific potential. As discussed in *Space Science in the 21st Century* (NRC, 1988), such an instrument could detect earth-like planets around nearby stars and perhaps detect O_3 at 10 μm or O_2 at 1 μm in their atmospheres. Oxygen molecules are believed to be universally related to the presence of life in an atmosphere like our own. This very large telescope could also study the formation and evolution of galaxies by taking images and spectra of galaxies at large redshifts. The question of what is the best location, a lunar base or high earth orbit, is particularly problematic for such a telescope. As for other cases, the answer will depend on technological developments in the next decade and on the infrastructure that will become available to support orbiting and lunar observatories.

Interferometry at Visible and Near-infrared Wavelengths

The techniques of interferometry can be used to link widely separated telescopes on the lunar surface to produce the spatial resolution of a single large telescope many kilometers in size. With an array of telescopes spread over a 10-km baseline, distortion-free images can be obtained for faint sources with

FIGURE 6.1 An artist's conception of an interferometer consisting of three telescopes operating at optical and infrared wavelengths on the moon. Light from the three telescopes is combined in the central building.

5- to 100-millionths-of-an-arcsecond resolution in the visible and near-infrared wavelengths, 0.2 to 5 μm. Astrometric observations could be made with high precision over this wavelength region.

Five passively cooled, 1.5-m telescopes operating together in this wavelength region could revolutionize many research topics in astronomy by combining high sensitivity with unprecedented spatial resolution (Figure 6.1). Such an instrument could map protoplanetary disks around young stars in the constellation of Taurus with a resolution better than 0.004 of the earth-sun separation, more than enough resolution to find gaps in the disks indicative of the presence of forming planets, to detect planets around stars out to a few thousand light-years by astrometric motions, to measure distances and motions of stars and star-forming regions in nearby galaxies, and to resolve the environment around the energy sources of quasars.

The full power of such an instrument would be difficult to realize on the earth because of the effects of the terrestrial atmosphere, and difficult to achieve in orbit because of the precise separations and orientations that would have to be maintained between component telescopes spread out over many kilometers. Several small, modular telescopes operating in concert could return fundamentally new astrophysical results relatively early in the life of a lunar base.

Interferometry at Submillimeter Wavelengths

The wavelengths between 100 μm and 1000 μm (1 mm) offer the key to many problems concerning the formation and evolution of stars and galaxies. Observations with infrared and radio techniques will determine the densities and temperatures in the protostellar nebulae of nearby star-forming regions and in distant star-burst galaxies, leading to a more detailed physical understanding of where and how stars form. The spectral line of ionized carbon at 158 μm is a fundamental cooling transition for star-burst and primordial galaxies and would be visible with a lunar observatory out to redshifts well beyond 3. An array of small, modular submillimeter telescopes spread over the lunar surface would allow thousandth-of-an-arcsecond imaging, adequate to search for evidence of planets forming in the disks surrounding nearby stars or to probe the energy sources of luminous infrared galaxies.

Radio Observations

The far side of the moon could provide a uniquely quiet environment for a large radio telescope after the initial development of a lunar base. It is important to preserve this area as a radio-free region in which especially sensitive scientific experiments could be performed.

High-Energy Astrophysics

The lunar surface might be suitable for the construction of x-ray and gamma-ray facilities that require large, stable structures, such as long-focal-length grazing-incidence telescopes, shadow cameras with large separations between mask and detector, or large detector arrays. Gamma-ray detectors could take additional advantage of the lunar soil for shielding against backgrounds from stray energetic particles and radiation.

AN EVOLUTIONARY PROGRAM OF TECHNOLOGICAL AND SCIENTIFIC DEVELOPMENT

The unique aspects of the lunar environment may lead to qualitatively new types of astronomical instruments and measurements, ones that are both technically and intellectually different from those possible with ground-based or orbiting telescopes. Current ideas and instruments give us only approximate guidance on what questions to ask. The lunar program should proceed as a step-by-step program with milestones of increasing scientific and technical scope. An early start on advanced technology development is essential.

As an example, consider one approach to the goal of a large multi-wavelength interferometer on the moon. The program could start by building interferometers on the earth and progress to earth-orbiting interferometers—equipment initiatives individually recommended in Chapter 1. Then, as the surface of the moon becomes accessible for astronomical facilities, small telescopes might first be installed and then, as experience is gained with the lunar environment, modest-sized interferometers could be constructed.

Such a step-by-step program applies equally to other types of telescopes and would have a number of advantages. The technology would be developed in a systematic way and realistically tested on astronomical sources at each step. For the specific example of the interferometer, different ideas for delay lines and beam combiners could be evaluated and the best technique chosen for the final multielement array. The scientific concepts would develop together in concert with the technology and could guide the choice of a suitable final instrument. Since no astronomical object has yet been observed with even thousandth-of-an-arcsecond resolution at optical wavelengths, entirely new phenomena are expected at the millionth-of-an-arcsecond level. Outstanding researchers would be attracted to the lunar initiative if they could foresee interim scientific results being obtained. Students could be involved in the intermediate stages of such a phased program. The opportunities for technological spinoff will be greater if there are intermediate goals that bring scientists and industry together frequently to make instruments that will be used for measurements of immediate scientific interest.

SPECIFIC TECHNOLOGY INITIATIVES

Many new technologies will be required for a lunar base. For example, robotic assembly of precision structures will be important for the construction of large telescopes. The ability to send scientific payloads to the moon independent of the system that transports humans may be critical for the long-term viability of such a program. These and other technological issues must be investigated thoroughly in this decade [see, for example, *Human Exploration of Space: A Review of NASA's 90-Day Study and Alternatives* (NRC, 1990b)].

Advanced technology development is needed to achieve the science described above. Ground-based and free-flying interferometric instruments need to be started within the decade of the 1990s in order to provide a basis for deciding whether it will be appropriate to place a major interferometric instrument on the moon. A large telescope operating from ultraviolet to infrared wavelengths, and placed either in earth orbit or on the moon, would be an instrument with immense power. Building such an instrument will require advances in making and supporting lightweight mirrors. It is necessary to begin required technology development soon. A submillimeter astronomy program in earth orbit is an appropriate preparation for an eventual lunar telescope. These technologies are also discussed in Chapter 1 in connection with the prioritized list of new technology initiatives for future NASA missions that may, or may not, be carried out on the moon.

THE IMPACT OF THE LUNAR PROGRAM

In large space projects, there is sometimes a temptation to move on to the next challenging program before exploiting fully the scientific potential of existing facilities. Some experiences of the research community with the Space Shuttle and the Space Station give cause for concern about possible future effects of a lunar initiative on NASA's scientific research efforts. A lunar base can be developed in a way that does not disrupt ongoing and planned programs. The *Strategic Plan* (NASA, 1988, 1989) for NASA's Office of Space Science and Applications (OSSA) describes a carefully balanced and scientifically important research program, many aspects of which have been reviewed and endorsed by this and other relevant National Research Council committees.

- **The committee recommends that NASA formulate its plans for a lunar initiative in a way that protects the base program in astrophysics from disruptions caused by problems in the Space Exploration Initiative.[1] The scientific potential of the ongoing and planned earth-orbiting observatories should be exploited before committing resources from the science budget to major lunar facilities that will not be operational until at least 2010. Initial funding for advanced technology development and for scientific missions that are precursors to lunar facilities should come from the Space Exploration Initiative.**

One of the key goals of the Space Exploration Initiative is to help interest young people in careers in science and engineering and thereby enhance the nation's capabilities in these areas. Astronomy can play an important role in inspiring a new generation of scientists if an adequate and stable fraction of the funding for the Space Exploration Initiative is dedicated to the development and execution of peer-reviewed scientific initiatives.

WHERE SHOULD THE PROGRAM BE IN 10 YEARS?

The Space Exploration Initiative is a multidecade program. The astronomical aspects of this program need a long-range plan that takes advantage

[1] This recommendation was also forcefully made in the report of the National Academy's Committee on Space Policy chaired by H.G. Stever (NAS-NAE, 1988).

of the unique properties of the moon and that follows an evolutionary path with appropriate milestones in planning, assessment of the lunar environment, development of technology and instruments, and scientific accomplishments. In keeping with the cycle of decennial surveys, it is reasonable to ask: where should the program be by the year 2001, at the time of the next decennial survey of astronomy and astrophysics?

By mid-decade, NASA should have completed analyses that will indicate which observations are best done from the moon. The analyses should consider the infrastructures, costs, risks, and environments of different sites, as well as possible benefits from advances in technology.

Key parameters of the lunar environment must be determined and site survey programs initiated. Flight hardware should be in development, or already in operation, by the end of the 1990s in order to answer questions that require in situ measurements. The earliest opportunity to obtain critical observations will be with the first Lunar Observer satellite. Any additional automated site survey missions should be under development by the middle of the 1990s.

The technology development programs in lightweight telescope construction, interferometry, and submillimeter techniques should be well under way. The first orbital precursor missions should be providing scientific data and technical experience as part of the Space Exploration Initiative. Possible programs that could be supported under this program include Delta-class satellites that would explore the feasibility of near-visible wavelength interferometry from space, of submillimeter astronomy with lightweight panels and cryogenic receivers, or of arrays of low-frequency dipole antennas. Programs in these or other appropriate disciplines in astronomy should be selected by peer review.

Preliminary studies have identified a few astronomical lunar facilities, such as a small (1-m-class) transit or pointed telescope, that might be appropriate early in a lunar program. At least one small project for the early phase of a lunar base should be selected by an open competition prior to the next decennial survey.

CONCLUSIONS AND RECOMMENDATIONS

The committee is convinced that the moon is potentially an excellent site for certain astronomical observatories that are capable of making fundamental discoveries. Operation from the moon may represent a significant advance over terrestrial or orbiting telescopes for interferometry at wavelengths ranging from the submillimeter to the optical.

- **The committee recommends that an appropriate fraction of the funding for a lunar initiative be devoted to fundamental scientific projects that will have a wide appeal, to supporting scientific missions as they progress from small ground-based instruments, to modest orbital experiments, and finally, to the placement of**

facilities on the moon. The advanced technology should be tested by obtaining scientific results at each stage of development.

- **NASA should initiate science and technology development so that facilities can be deployed as soon as possible in the lunar program. The NASA office responsible for space exploration and technology should support the long-term development of technologies suitable for possible lunar observatories.**

- **Site survey observations from the Lunar Observer(s), possibly with soft landed experiments such as a small transit telescope, should be a high priority for a lunar program. The requirements for astronomical observations should be carefully considered in the selection of the site for a lunar base.**

Multiwavelength (ultraviolet to infrared) observations with a large (16-m-class) telescope and infrared observations with a large, cold infrared telescope in a polar crater, or radio observations from the far side of the moon could offer unprecedented capabilities for astronomy. These projects are, however, formidable technical challenges.

- **NASA should develop the technology necessary for constructing large telescopes and should investigate which of these facilities are best placed in earth orbit and which are best placed on the moon.**

- **NASA, along with other governmental and international agencies, should strive to have the far side of the moon declared a radio-quiet zone.**

7

Policy Opportunities

INTRODUCTION

Federal agencies have invested in astronomical facilities in space, on the ground, and even underground to take advantage of technological advances leading to unique opportunities for research into many aspects of the universe. There are critical issues associated with these investments that must be addressed by the astronomical community and the sponsoring governmental agencies. How can astronomers, returning value to the nation, contribute to the national effort to improve science education and to stimulate interest in science or engineering as an attractive career? What is the proper balance between construction of new facilities, the maintenance and refurbishment of existing facilities, and support of individual researchers? What are the roles of large space missions, such as the Great Observatories, and of smaller, more frequent missions? What can be done to shorten the time between conception and completion of space programs? Under what circumstances should the United States carry out a project with international partners, and what kinds of projects are best carried out with national resources?

Before addressing these difficult problems, the committee first outlines some of the agency-related activities in astronomy in the 1980s. It then describes paramount concerns for the 1990s: an educational initiative in astronomy, the urgent need to revive ground-based astronomy, the opportunities and frustrations of space astrophysics, and the circumstances under which international collaborations are most fruitful.

THE PREVIOUS DECADE

During the 1980s, Congress and the relevant federal agencies responded positively to the advice offered by the previous Astronomy Survey Committee in the "Field Report" (NRC, 1982). The success in implementing recommended programs despite limited resources was possible in large part because of the work of dedicated people in government service.

The **National Science Foundation (NSF)** implemented, fully or partially, several of the Field Committee recommendations for new ground-based facilities. For example, the Very Long Baseline Array (VLBA) for radio astronomy is now nearly completed. The NSF also provided partial support to build two new 4-m-class telescopes at universities and initiated design work on 8-m-class telescopes. The NSF also supported the construction of a submillimeter telescope on Mauna Kea.

In addition, the NSF Astronomy Division responded to the astronomical community's enthusiasm in initiating the solar Global Oscillations Network Group (GONG) project. Through the Division of Polar Programs, the NSF supported an innovative research program that exploits the unique advantages of the South Pole for astronomy, and through the Physics Division, NSF carried out major programs in particle astrophysics such as the Fly's Eye telescope, the Chicago Airshower Array (CASA) project, and theoretical investigations.

Despite the scientific and technical accomplishments of the past decade, a major crisis has developed in the support of ground-based astronomy. The number of observers doubled, and major new observational facilities were added, because fundamental scientific problems were ripe for solution and required ground-based facilities. Nevertheless the support for facilities and basic research has decreased in purchasing power to 36 percent of what it was, per astronomer, in 1970, resulting in a serious erosion of the research infrastructure.

The policy framework for the **National Aeronautics and Space Administration's (NASA)** space astronomy and astrophysics program was provided by the *Strategic Plan* (NASA, 1988, 1989) for the Office of Space Science and Applications. The strategic plan strongly supported the four Great Observatories recommended by the Field Committee in 1982. The Hubble Space Telescope (HST) is operating, and instruments that will correct for the flawed mirror are expected to be available in a few years. The Gamma Ray Observatory (GRO) is scheduled to be launched in 1991. The Advanced X-ray Astrophysics Facility (AXAF) is under construction, with final approval contingent on successful tests of the mirrors. NASA's goal of providing new windows to the universe, if successfully completed, will be among the most important organized intellectual efforts of the 20th century.

The second-ranked intermediate program of the Field Committee, the Far Ultraviolet Spectroscopy Explorer (FUSE), is now under development, and the highest-priority moderate new initiative of this committee is to provide an independent spacecraft for FUSE. The overall budget for Explorer satellites was increased in the 1980s, although fewer Explorer missions were launched then than in the 1970s. Small Explorer (SMEX) missions, and a continuing suborbital program including rockets, balloons, and aircraft, made possible important scientific advances in the 1980s. NASA's Astrophysics Division supported important international collaborations on European, Japanese, and Soviet missions and began modernizing the university infrastructure necessary for research. The growth of the NASA grants program during the 1980s tracked the overall growth of the number of astronomers (Appendix B) and is a positive sign of NASA's willingness to support the research activities of scientists interested in data from space missions.

While maintaining primary responsibilities for ground-based and space astronomy, respectively, NSF and NASA sometimes worked together effectively during the 1980s toward the common goal of understanding the universe. Both agencies recognized that complementary data obtained from space and from the ground are essential for the solution of many important scientific problems. NSF's distribution to ground observatories of NASA-developed advanced detectors is a particularly successful example of this cooperation.

Despite the major advances of the 1980s, the decade was also a period of frustration for astronomers, NASA, and the nation. Only 2 American astronomical satellites were launched in the 1980s, compared to 10 in the 1970s. The time for development of a space mission has stretched to more than a decade and seriously interferes with the productivity of missions. These problems are due in part to NASA's reliance, in the early 1980s, on the Space Shuttle as a launch vehicle for astrophysics missions, a policy that has since changed. From 1981 to 1989, the average rate of Shuttle launches was about four per year, which was not adequate to accomplish all that the Shuttle was designed to do.

The greatest scientific disappointment to astronomers, and to the nation, was caused by the discovery in June 1990 of the flaw in the HST mirror produced almost a decade earlier. The committee discussed, in the brief interval between this writing and the revelation of the mirror's imperfection, ways that astronomers could try to help prevent similar disasters in the future. In the "Balanced Space Astrophysics Program" section below, the committee comments on some strategies that may reduce technical risks and make NASA's space astronomy program more efficient.

Scientists supported by the **Department of Energy (DOE)** at universities or at DOE's national laboratories have performed pioneering theoretical research and made important astronomical discoveries for more than 25 years. With DOE support, the field of observational neutrino astronomy was founded, and DOE continues to play a crucial role in this research area. DOE scientists at Lawrence Livermore and Los Alamos National Laboratories have been leaders in calculations of gravitational collapse, supernova explosions, stellar pulsations, nucleosynthesis, equations of state of dense matter, and stellar opacity, as well as in observing x-ray and gamma-ray sources.

The DOE appreciates the interaction between different areas of fundamental research and the impossibility of knowing a priori the directions in which pure research will lead. As a response to a number of scientific developments in the 1980s, the Directorate of High-Energy and Nuclear Physics has informed this committee that in the 1990s it will consider supporting astrophysical research that is related to its mission of seeking a deeper understanding of the nature of matter and energy and the basic forces that exist between the fundamental constituents of matter.

Some of the basic research and technology programs at the **Department of Defense (DOD)** make essential contributions to astronomical research. Examples include astrometry and optical interferometry at the U.S. Naval Observatory, development of space instrumentation by the Naval Research Laboratory, innovations in infrared detector technology by the Air Force Office of Scientific Research, and cryogenic and adaptive optics technologies developed as a result of the Strategic Defense Initiative. New opportunities exist for synergism between astronomical research and the nation's defense needs. The committee believes that these opportunities should be exploited.

EDUCATIONAL INITIATIVE

The education of young people is the foundation for future scientific and technical advances. Thus the committee begins its discussion of policy issues with a discussion of an educational initiative in astronomy.

The nation's colleges and universities are training too few Americans in science, engineering, and mathematics (*A Nation at Risk;* NCEE, 1983). In a world in which technical skills and quantitative reasoning are increasingly important, the nation needs more individuals with scientific knowledge in order to improve the quality of daily life and to help secure our economic competitiveness. Too few American students enter college with an adequate background in science and mathematics and with the intention of pursuing scientific careers. Of those entering college with an initial interest in science, too many ultimately obtain degrees in other areas, exacerbating the problem. Unless current trends are reversed, our nation will soon suffer a critical shortage of trained individuals who can take advantage of opportunities for scientific discovery or for technical innovation. As astronomers, the committee is committed to helping solve this national problem.

Television and the popular press expose young people to many challenges in business, law, and medicine but usually fail to present the exciting opportunities in science and technology. As discussed in Chapter 8, astronomy has a special appeal to young people and is particularly effective in stimulating interest in science and engineering at an early age.

The committee emphasizes below programs relating to precollege education. Several additional proposals have been described in the document *An Educational Initiative in Astronomy* (Brown, 1990) and in the study by the Policy Panel in the *Working Papers* (NRC, 1991) of this report.

- **The committee recommends that NSF establish, at one or more of the major U.S. observatories, an office for astronomical education with responsibility for involving professional research astronomers in educational activities, for making available material about astronomy, for assisting with teacher workshops, for promoting student involvement in research, and for providing guidance on curriculum matters.**

The education offices at NASA centers are doing an excellent job with limited resources and should be strengthened.

- **The committee recommends the expansion of summer programs and workshops at universities and national research centers for paid in-service training of science teachers.**

Such workshops provide excellent opportunities for science teachers to gain direct experience with modern astronomical research and to make contacts with astronomers who are committed to improving science education. Workshops are particularly effective when they attract master teachers who are developing curriculum materials and training other teachers.

- **The committee recommends that NSF establish a national Astronomy Fellowship program that will allow each state to select an outstanding high school student as a state fellow in astronomy.**

The state fellows would serve as paid science interns during the summer months at one of the major national or private observatories, where they would participate as assistants in the research of the professional staff. The program would show young people that a career in science is feasible and exciting. The committee suggests that one of the national astronomy research centers act, in cooperation with the appropriate agencies and other major astronomical institutions, as the organizer and coordinator of the Astronomy Fellowship program.

- **The committee recommends that the American Astronomical Society establish an annual prize in recognition of outstanding contributions to secondary or college science education.**

The educational programs in astronomy should be a joint effort involving both the educational and the research branches of the relevant agencies. The educational directorates, which would supply the primary funding for these activities, can help researchers bring the excitement of modern astronomy into the nation's classrooms.

Adequate public education at all grade levels from kindergarten to college, starting with basic numeracy and literacy and ending with a solid grounding in humanistic and scientific concepts, is a long-term solution to the numerical imbalances in racial, ethnic, and sexual representation in different fields of science, including astronomy (Appendix B). In addition to the recommendations made above, the committee urges astronomers to take personal action to improve science education in their communities through presentations at local schools and by visits of students to nearby astronomical facilities.

As noted in Appendix B, there are many more active astronomers than there are faculty positions. Establishing additional faculty positions would bring more of the excitement of astronomy directly to students. The committee commends NASA's attempts to work with universities to provide additional tenure-track positions in astronomy for those young astronomers interested in teaching.

REVIVING GROUND-BASED ASTRONOMY

Ground-based astronomy is imperiled by inadequate funding and the consequent deterioration of major facilities and loss of key personnel. Without adequate support for ground-based work, the United States will lose many of the fruits of both the space and the ground astronomy programs.

The current crisis in the U.S. ground-based program is due to the long-term funding history of the NSF, which has held astronomy to essentially the same base budget (in Consumer Price Index-adjusted dollars) for the past 20 years. In response to the explosive growth in scientific opportunities, new facilities have been constructed and the number of astronomers using ground-based facilities has doubled since 1970, but funding to operate and maintain these facilities and to conduct basic research has been constant. The National Optical Astronomy Observatories (NOAO) has opened two new 4-m optical telescopes and has absorbed the operation of Sacramento Peak Solar Observatory. The National Radio Astronomy Observatory (NRAO) has opened the Very Large Array (VLA) and has begun to operate the VLBA network. At the same time, improvements in astronomical instrumentation have greatly increased the capabilities of optical and radio telescopes, and the total number of visiting observers at all national observatory sites has tripled. Despite these increased responsibilities, staffs at the national observatories have been cut. Because of funding constraints, available instrumentation and computing systems lag behind the state of the art, precious data have been inadequately analyzed, and expensive equipment has been poorly maintained.

The lack of adequate NSF funding for grants and for the infrastructure has created serious problems throughout ground-based astronomy; some examples are listed below.

The success rate for first-time proposals from young scientists has fallen to 1 in 10. The purchasing power of the average grant, and the per capita funding available to astronomers, have both dropped by about a factor of 2 since 1980 (Appendix B). The number of postdoctoral positions supported by grants has fallen by 20 percent. The operations budgets of the NRAO, NOAO, and the National Astronomy and Ionosphere Center (NAIC) dropped by 20 to 35 percent in the 1980s. The staffing levels at the national observatories dropped by more than 15 percent in the 1980s. Cuts at NOAO led to closure of a heavily used 36-in. telescope, suspension of travel support for observers going to the Kitt Peak (Arizona) and Cerro Tololo (Chile) telescopes, and closing of the advanced projects group responsible for the development of innovative optics concepts such as adaptive optics and interferometry. Cuts at NRAO led to deferred maintenance that reduced the efficiency of the VLA, a lack of the computer resources needed to process spectral-line data at the VLA, and an extended and more expensive construction schedule for the VLBA.

The committee has expressed its view in Chapter 1 that the highest funding priority for the 1990s in ground-based astronomy is restoring support for the scientific infrastructure, especially for grants to individual researchers and for maintenance and refurbishment of frontier national facilities.

The number of U.S. astronomers has increased by about 40 percent over the last decade (Appendix B). New PhDs account for only about half of the new astronomers; the remainder have moved into astronomy from other fields. The increase in the number of professional astronomers indicates the intellectual excitement that astronomy provides for students and for scientists in other fields.

The committee judges that the number of astronomers and their current rate of production is appropriate to the capital investment being made by the United States in new telescopes. NASA has estimated that a pool of astronomers at least as large as the present research community, or perhaps slightly larger, will be required to analyze the important data that will be returned from its space missions over the next 10 years.

BALANCED SPACE ASTROPHYSICS PROGRAM

The *Strategic Plan* (NASA, 1988, 1989) developed for NASA's Office of Space Science and Applications incorporates the unfinished astrophysics missions, recommended by the astronomical community, that were selected and started in the 1970s and 1980s. The completion of this strategic plan will determine most of the astrophysics missions to be launched until the late 1990s. The committee endorses this plan. The recommendations made here are intended primarily to affect the process by which NASA will select and carry out the astrophysics missions to be started during the latter half of the 1990s and beyond.

The success of the U.S. space astrophysics program depends on a proper balance between large, moderate, and small missions. Large missions such as the Great Observatories have capabilities that cannot be matched by smaller missions. They provide leaps in capability needed to solve many of the problems at the frontiers of the universe. Large missions involve many researchers in innovative instrument development, support broad community participation in creative observing, attract students, and capture the public imagination. For example, data from the Einstein Observatory have been analyzed in more than 1,000 published papers, nearly all of which were written by individuals or by small groups of investigators. Large missions in astronomy provide for "small science with big facilities." Amortized over their lifetimes, large missions can be efficient and cost-effective.

However, large missions are also complex and expensive. NASA and the scientific community must be alert to possible technical and management simplifications to assure scientific success on the fastest schedule and at the lowest cost. The manufacturing flaw in the HST mirrors constitutes a sober lesson, but HST problems must be viewed alongside a list of stunning successes in other complex missions such as the Viking and Voyager planetary flybys and the High-Energy Astronomical Observatory (HEAO) program, including the Einstein Observatory. The astronomical community and NASA must use the lessons of HST, and of other complex NASA science projects, to learn how to improve the management of future large missions.

At the same time, moderate and small missions add a vital dimension to NASA's space science program: the ability to deploy new instrumental technology into space on relatively short time scales. Smaller programs can provide new approaches to well-defined scientific problems of great significance that are not easily addressed with the large missions. The opportunity for rapid access to space allows for quick responses to scientific and technical developments, stimulates progress in technology, and attracts young instrumentalists who are essential for a successful future in space science. Many of the outstanding astronomical missions of the past were of moderate or small size, such as the Uhuru x-ray telescope, the International Ultraviolet Explorer (IUE), the Infrared Astronomical Satellite (IRAS), and the Cosmic Background Explorer (COBE) satellite. The small suborbital program made major contributions to the study of Supernova 1987A.

- **The committee recommends that NASA continue to develop a vigorous program of moderate and small missions of limited complexity and shorter development times, with increased use of expendable launch vehicles.**

Recent NASA Announcements of Opportunity have stimulated many innovative proposals for scientific payloads for small and moderate missions. In addition, exciting astrophysics missions currently under development by Japanese and European space agencies will use mission concepts and instrumental technologies that were invented and developed by U.S. astrophysicists. A number of Explorer-level proposals of great scientific importance are described in Chapter 1 and in the panel reports contained in the *Working Papers* (NRC, 1991) of this report. NASA has recognized the need for a more frequent launch rate of small and moderate-sized astrophysics missions and has begun to respond to this need within its *Strategic Plan* (NASA, 1988, 1989).

- **The committee recommends that NASA increase the rate of Explorer missions for astronomy and astrophysics to six Delta-class and five SMEX missions per decade.**

The committee believes that the most successful and cost-effective projects, large, moderate, and small, involve an intimate partnership between university scientists and the NASA and industrial communities. For large programs, active participation by NASA and non-NASA scientists at the project level facilitates the most cost-effective allocation of resources and helps ensure that scientific objectives are met. The committee believes that the most qualified astronomers should assume major responsibilities for important projects and that large programs are more likely to succeed if one accomplished individual has full knowledge and appropriate responsibility for each project. In smaller programs, the shorter schedules and more limited budgets require still closer ties among the different participants in the program. National Research Council studies of the Explorer program (NRC, 1984, 1986b) have called for simplified project control of moderate and small missions, including clear authority for a scientific principal investigator. Specific proposals for ways to make moderate and small missions more effective are also contained in the *Working Papers* (NRC, 1991). NASA is exploring such approaches in several small missions.

- **The committee believes that NASA should increase the role of scientists in the management of large, moderate, and small projects.**

Long development times and high costs for small and moderate missions have limited in recent years the effectiveness of the Explorer program. In addition to the management issues just cited, the National Research Council studies of the Explorer program called for a return to missions with well-defined scientific objectives, careful attention to mission cost in the selection process, and appropriate levels of formal requirements for reliability and quality assurance. The High-Energy Transit Experiment (HETE; a shuttle-launched satellite package for detecting gamma rays) is being developed in this way by NASA though a grant to university scientists. NASA is also moving to implement these recommendations in the SMEX program.

- **The committee urges NASA to continue streamlining management practices to assure well-defined science objectives, accurate cost control in the selection process, and appropriate requirements for reliability and quality assurance.**

The committee is worried that the operation of the next three Delta-class Explorer missions is currently based on a single, reusable spacecraft called the Explorer platform. The current plan is to launch the Extreme Ultraviolet Explorer (EUVE) attached to the Explorer platform with a Delta rocket. Subsequent Explorers would be launched with Space Shuttle missions during which astronauts would exchange EUVE (or its successor) for a different observatory to be operated on the orbiting platform. The committee believes that this serial approach involves a significant risk for long and expensive delays, reduces mission lifetime, and restricts projects to low-earth, Shuttle-accessible orbits. The committee's highest-priority moderate space initiative is, as discussed in Chapter 1, an independent spacecraft for FUSE, which would obviate dependence on the single, reusable Explorer platform and would enhance science opportunities for the FUSE mission. The committee notes that a previous endorsement of refurbishable shuttle-serviced spacecraft by the NRC's Committee on Space Astronomy and Astrophysics (NRC, 1986b) was made before the full implications of the Challenger tragedy were recognized.

- **The committee believes that there are scientific, programmatic, and financial advantages to using independent spacecraft, versus a reusable shuttle-serviced platform, for Explorer missions.**

INTERNATIONAL COOPERATION

Research in astronomy and astrophysics is an international enterprise. Recent examples of successful international collaborations include the GONG project for studying the motions of the sun's surface, the IUE and IRAS satellites, and intercontinental radio interferometry, which maps distant quasars and the motions of terrestrial continents. The United States provides access on a competitive basis for scientists of all countries to a number of major U.S. national facilities. In turn, the United States should expect that other countries will provide reciprocal access to their major facilities.

International cooperation in building major facilities is most effective when the project draws on the complementary capabilities of different nations, or when the projects are too expensive for individual nations to afford. Some projects, such as a permanent observatory on the moon, are so large and complex that international collaboration may be essential.

Substantial extra costs can be incurred, however, when facilities are built by more than one nation, arising, for example, from the increased complexity of coordinating technical interfaces and the necessity for duplicating some administrative efforts. Sometimes, international collaboration and scientific goals are most effectively advanced when nations build their own unique facilities, providing access to qualified scientists from other nations.

- **The committee recommends that international cooperation be considered for the development of a major initiative if such a project draws on complementary capabilities of different nations or requires resources beyond those that can be provided by the United States alone.**

8
Astronomy as a National Asset

Astronomy makes humanistic, educational, and technical contributions to our society. The most basic contribution of astronomy is that this science provides modern answers to questions about humanity's place in the universe. We can now give quantitative answers to questions about which ancient philosophers could only speculate. In addition to satisfying our curiosity about the universe, astronomy nourishes a scientific outlook in society at large. Society invests in astronomical research and receives an important dividend in the form of education, both formally through instruction in schools, colleges, and universities, and more informally through television programs, popular books and magazines, and planetarium shows. Astronomy introduces young people to quantitative reasoning and helps attract them to scientific or technical careers. Modern astrophysics also contributes to areas of more immediate practicality, including industry, medicine, and defense.

OUR PLACE IN THE UNIVERSE

As far back as history records, we humans have attempted to understand the origins of the universe and our place in it. The Copernican revolution showed that the earth was not the center of the universe, but rather only one of several planets orbiting the sun. With further study, our sun was recognized as a typical star, located far from the center of a normal galaxy. Astrophysical measurements of the motions of binary stars and of the atomic transitions of various elements in nearby stars and distant galaxies have demonstrated that the laws of physics are the same at distant times and places as here and now on the earth. Modern cosmological models suggest that the universe is only three or four times older than the earth itself and that the universe has neither center nor edge.

Astronomy helps reveal the nature of life and its fragility. Our knowledge of the formation of the elements tells us that we are "stardust," formed from material built up in the early seconds of the universe or in the interiors of massive stars. Chemical reactions of carbon atoms with other atoms in interstellar gas and in meteorites produce the same molecules that are the building blocks of living creatures on the earth. Looking down at the earth from orbit, we see our home as a small, fragile entity. Observing the inhospitable environments of Mars and Venus lends reality to concerns for the future of our own ecosystem.

As new instruments expand our capability to peer deeper into space, our view of the universe expands and evolves. Astronomers know that although any given interpretation may be outmoded by a new observation, the basic process of empirical inquiry is sound and leads to a better understanding of the world around us. Astronomers can offer to their fellow citizens the confidence that the universe is comprehensible.

ASTRONOMY AND AMERICA'S SCIENTIFIC LEADERSHIP

Public Scientific Literacy

Reports such as *A Nation at Risk* (NCEE, 1983), *A Challenge of Numbers: People in the Mathematical Sciences* (NRC, 1990a), and *Physics Through the 1990s* (NRC, 1986a) have called for improvements in public education at all levels, and particularly in science. Astronomy and astrophysics have an important role to play in maintaining and restoring American leadership in science and technology by raising the level of scientific literacy among the general public and students at all levels, by inspiring students to become scientists, and by training scientists for other technical careers.

Astronomical concepts are usually included as part of the physical sciences courses taken by elementary and junior high school students and occasionally appear as parts of high school curricula. Project STAR (Science Through its Astronomical Roots) is being developed at the Center for Astrophysics to provide astronomical course material for high school chemistry and physics classes. Project 2061, which aims to ensure scientific literacy among all high school graduates by the year 2061, when Halley's Comet returns, has a significant astronomical component. Astronomy is also the focus of many successful adult education courses and teacher training programs.

Formal astronomy courses probably have their greatest impact at the non-major undergraduate level. Colleges and universities with astronomy (or physics and astronomy) departments had 1.2 million undergraduates in 1988;

about 8 percent of these students took introductory astronomy (Sadler et al., 1989). Large numbers of students also learn about science from astronomy courses given in institutions without formal astronomy departments.

Cosmos, the most successful experiment ever in public scientific education, has had an audience of 400 million television viewers in about 60 countries. The book *Cosmos* (Sagan, 1983) is the best selling English-language science book in history. But *Cosmos* is only one of the ways that astronomy is presented to the public. *Project Universe*, a series of 30 half-hour programs, had its hundredth showing last year. *NOVA* often features astronomical subjects. The daily radio feature "Stardate," produced at the University of Texas, is broadcast by about 200 stations and has attracted half a million letters from listeners in the past decade.

While astronomers make up only 0.5 percent of the scientists in the United States, science magazines such as *Scientific American, Discover*, and *Science Digest* devote about 7 percent of their pages to astronomy. Over the last 10 years, astronomy articles outnumbered those on biology and nearly equaled the number of articles on physics in some of the most prestigious newspapers. Of the 10 best-selling nonfiction books on the *New York Times* list in 1988, three were on astronomy. The intellectually demanding *A Brief History of Time* (Hawking, 1988) spent over two years on the *New York Times* best seller book list. About 250,000 enthusiasts subscribe to *Sky and Telescope* or *Astronomy* magazine.

Planetariums and observatories reach millions of children and adults. McDonald, Palomar, and Kitt Peak Observatories each record almost 100,000 visitors per year. The Griffith Observatory and its associated planetarium in the hills above Los Angeles hosted 1.7 million people in 1988, as many as the Los Angeles County Museum of Art and the J.P. Getty Museum together.

Training of Professional Scientists

A steadily decreasing number of people in the United States are pursuing careers in science and engineering. Our economy depends on our ability to compete technologically with other nations. The quality of the environment depends on developing safe, clean industries and sources of energy. None of this can be accomplished without a work force of imaginative and highly trained scientists and engineers. In what ways can astronomy contribute?

First, because of its broad appeal, astronomy is often the science that initially arouses the scientific interests of people who eventually earn degrees in other technical disciplines. For example, two long-standing summer programs in astronomy for outstanding high school students, at the University of Illinois and at the Thacher School in Ojai, California, have inspired nearly all their participants to go on to college programs in science, engineering, mathematics, or medicine. Chapter 7 makes specific recommendations about enhancing the role of astronomy in science education at the precollege and college level.

About 70 American colleges and universities currently offer degrees in astronomy, astrophysics, or closely related fields. Recipients of bachelor's degrees in astronomy are widely diffused in technical fields. Surveys of students (Ellis and Mulvey, 1989) suggest that about half go on to graduate school in physics, engineering, geology, and atmospheric science as well as astronomy (Figure 8.1); the other half are employed mostly in technical industrial firms.

Most recipients of graduate degrees in astronomy establish lifetime careers in astronomical research and teaching. However, surveys conducted at the California Institute of Technology and the University of Maryland suggest that about 40 percent of those who earn doctorates in astronomy eventually move into research and teaching in other fields, or take jobs in government, industry, or defense.

A graduate degree in astrophysics provides a good match to the requirements of scientific defense work. About 100 members of the American Astronomical Society are currently employed at Los Alamos and Lawrence Livermore National Laboratories; an even larger number are engaged in full- or part-time defense work elsewhere.

SYNERGISM WITH OTHER SCIENCES

The universe is a laboratory far grander than any constructed on the earth. The extreme conditions reached in astronomical systems, of high and low temperatures, of high and low densities, of common and rare elements, provide unique tests of physical theories.

High-Energy and Particle Physics

Astronomical observations are helping to clarify the properties of the neutrino, an elusive elementary particle whose nature is poorly understood. For example, the best limit on the charge of the neutrino is based on observations of the neutrino burst associated with Supernova 1987A. One of the most intriguing problems in high-energy physics originated in an attempt to use neutrinos to look into the nuclear-burning core of the sun and thereby test directly the theories of stellar evolution and nuclear energy generation in stars. As a result of the discrepancy that arose when observations disagreed with predictions, many new testable ideas regarding neutrinos were developed. While some of these theories proposed changes to models of the solar interior, others have raised questions about the fundamental physics of the neutrino. The explanation of the conflict between observation and theory for solar neutrinos may ultimately provide evidence for lepton nonconservation and for at least one neutrino having a non-zero mass.

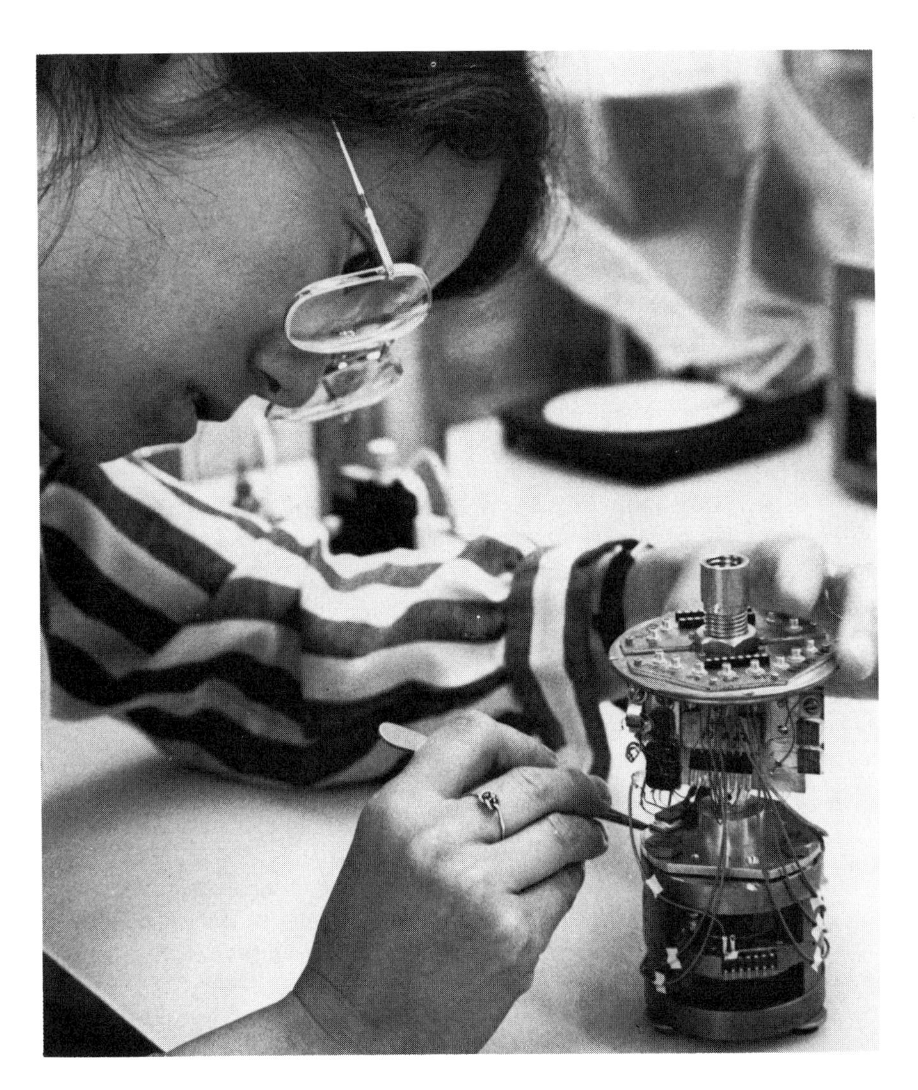

FIGURE 8.1 The training of astrophysicists can include building advanced equipment. Here Berkeley graduate student Ning Wang assembles a prototype of a cryogenic detector being built for the search for weakly interacting dark matter particles. Photograph courtesy of Technical Information Department, Lawrence Berkeley Laboratory, University of California.

Geophysics

Techniques from radio astronomy are being used to measure the motion of the earth's continents. Very long baseline interferometry (VLBI), normally used to measure the positions of celestial radio sources, also measures the position of radio telescopes to a few centimeters. VLBI has determined that the rate of slippage along the San Andreas fault, and the rate at which the Atlantic Ocean is widening, is from 1 to 3 cm per year.

Astronomers also study the geology of other planets, thereby giving geologists a better perspective on terrestrial problems. Examples include insights into vulcanism and tectonic mechanisms obtained by comparing Earth to Mars, Venus, Io, and Triton.

ASTRONOMY AND THE EARTH'S ENVIRONMENT

An Astronomical Context for the Earth's Environment

The energy from the sun is a critical parameter determining the habitability of the earth. There are tantalizing hints of terrestrial climatic changes on long time scales driven by changes in the solar luminosity and in the earth's orbit around the sun. We need to understand solar effects on our environment in order to isolate and better understand human perturbations on the environment. Measurements from space during the most recent solar cycle show that the total brightness of the sun changes about 0.1 percent between maximum and minimum activity (Figure 8.2). Most models suggest that such changes should cause a minor global temperature change of less than 0.1°C. However, the integrated solar activity, smoothed over several cycles, has increased roughly in phase with the apparent 0.5°C warming trend of the past century; the Little Ice Age in the 1600s coincided with an extended period of exceptionally low solar sunspot activity. We need to understand more about climatic cycles on both the sun and the earth to determine how significant these effects are. A promising way of understanding solar cycles is to compare the sun with the other stars known to have 5- to 20-year cycles.

There are numerous short-term effects of sudden increases in the ultraviolet, x-ray, and particle radiation coming from the sun. The high-energy radiation from such flares hits the earth's upper atmosphere, causing heating and ionization, and also modifies terrestrial electric and magnetic field structures on the ground. The 11-year cycle of solar activity reached near-record levels in 1989. Documented effects of large solar flares include major power outages, disruption of radio communication, and added drag on satellites. Continuous monitoring of the sun can give a few hours warning of the arrival of particles from solar flares. These observations will be important for the safety of astronauts inhabiting, or traveling to, the moon or Mars.

Models of the Earth's Environment

Questions of long-term climate change and the influence of humans on the earth's environment are of concern to citizens and governments. Much of the research on which the discussion is based originated in attempts to explain the climates of other planets, including Venus with its runaway greenhouse effect, Mars with its thin atmosphere, and Jupiter, Saturn, and Neptune with their dramatic storm systems (Plate 8.1). Astronomers and atmospheric scientists have developed models to help understand and eventually predict the dynamics of planetary atmospheres and the physical conditions that result in environments hospitable to life. Some of the models work tolerably well for the simplest planetary atmosphere, that of the planet Mars, but fail for more complex planets, including our own. These models can be improved by comparison with the observations of other planets to make reliable predictions for our own environment.

Astronomy, Weather, and Ozone Depletion

Weather satellites are one of the practical benefits of the space age. The tropospheric temperature sounders used in national security applications and soon to be used in civilian weather satellites are direct descendants of the planetary radio astronomy instruments used to probe the atmosphere of Venus. Remote sensing from satellites is one of the best methods for monitoring the earth's ecosystem.

Radio astronomers, for example, have adapted the techniques of millimeter wave astronomy to studies of ozone depletion. In 1977 astronomers initiated a program to measure the stratospheric concentration of chlorine oxide (ClO), the most important tracer of the destruction of ozone by chlorofluorocarbons. Measurements of the diurnal variation of ClO in the middle stratosphere provided a critical test of the proposed photochemical models (Solomon et al., 1984). Subsequent measurements of the high concentration of ClO in the lower stratosphere during early spring and its subsequent disappearance in October (de Zafra et al., 1987), when ozone levels returned to normal, demonstrated that chlorine chemistry was responsible for the Antarctic ozone hole (Plate 8.2). Automatic millimeter wave instruments built by a commercial company founded by astronomers will measure both ozone and ClO as part of a worldwide network.

Atmospheric ozone also varies significantly due to natural causes. The solar activity cycle produces an 11-year variation in the sun's ultraviolet radiation, and this in turn affects the terrestrial ozone abundance. Solar variability in the ultraviolet must be known in sufficient detail to delimit the natural causes of ozone change before one can confidently extract the man-made component of that change. Ultraviolet solar variability is thus of practical as well as astronomical interest.

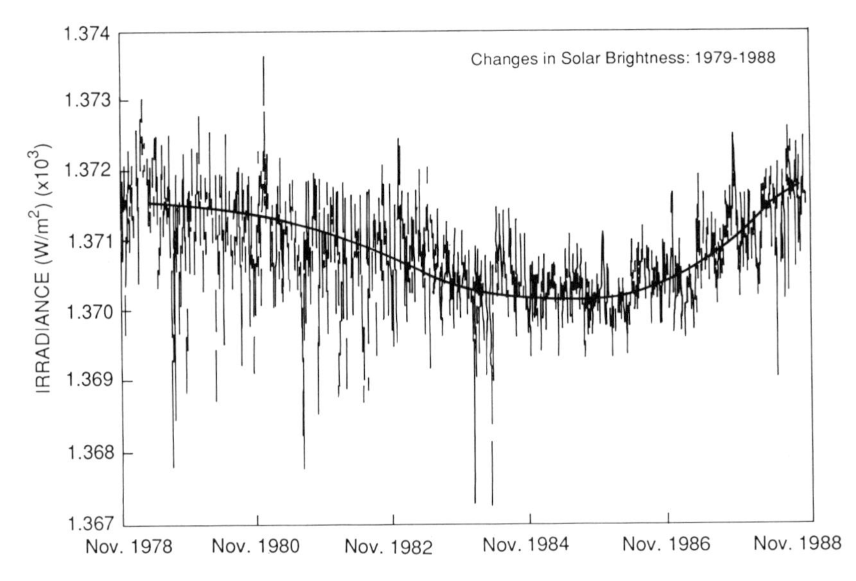
Changes in Solar Brightness: 1979-1988
IRRADIANCE (W/m²) (x10³)
1.374
1.373
1.372
1.371
1.370
1.369
1.368
1.367
Nov. 1978
Nov. 1980
Nov. 1982
Nov. 1984
Nov. 1986
Nov. 1988

a

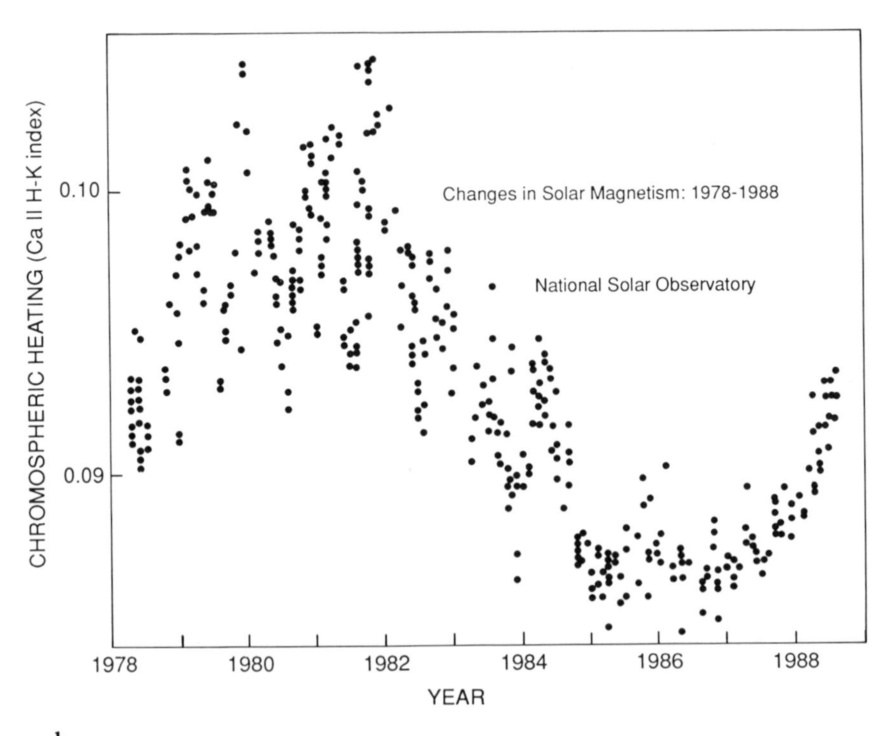
Changes in Solar Magnetism: 1978-1988
National Solar Observatory
CHROMOSPHERIC HEATING (Ca II H-K index)
0.10
0.09
1978
1980
1982
1984
1986
1988
YEAR

b

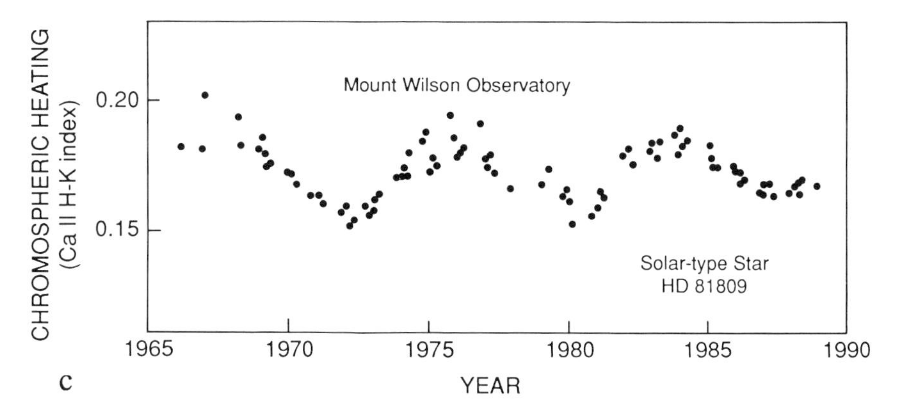

FIGURE 8.2 Instruments on the Solar Maximum Mission and Nimbus-7 satellites measured a small-amplitude (0.1 percent) variation in the total energy output of the sun during the past 11-year activity cycle (a, facing page). These variations are associated with measured changes of heating in the sun's atmosphere (b, facing page), which also can be seen in other stars like the sun (c, above). The study of stars like the sun gives us more data on these variations in luminosity and provides a context in which to understand long-term variations in the terrestrial climate. Figure (a) reprinted by permission from Hickey et al. (1988), copyright © 1988 by Pergamon Press; figure (b) courtesy of W. Livingston and O.R. White, National Solar Observatory; and figure (c) courtesy of S. Baliunas, Mt. Wilson Observatory.

USES OF ASTRONOMICAL TECHNIQUES OUTSIDE ASTRONOMY

Useful applications often arise when scientists develop new research techniques. A few examples of the application of astronomical techniques outside astronomy are described below; more are cited in Tables 8.1 to 8.4. Astronomers are currently working on infrared imaging devices suitable for low light levels, special-purpose computers for following simultaneously the motions of thousands of particles, and low-noise radio receivers for submillimeter wavelengths. No one can state for certain what these new devices will be good for, but as Michael Faraday replied, when questioned by the then British Chancellor of the Exchequer William Gladstone about the utility of his new theories of electricity, "Why, Sir, there is every probability that you will soon be able to tax it."

Medicine

Medicine and astronomy share the problem of imaging the inaccessible. Astronomers, especially radio astronomers, led the way in solving the general problem of reconstructing the two- or three-dimensional appearance of objects from a number of one- or two-dimensional scans (Table 8.1). A paper by Bracewell and Riddle (1967) is widely cited in the nonastronomical literature. Some of the image-reconstruction techniques of radio astronomy are now used

TABLE 8.1 Medical Applications of Astronomical Techniques

Astronomical Technique or Device	Medical Uses
Image reconstruction from one- and two-dimensional scans of radio sources	Imaging for CAT scans Magnetic resonance imaging Positron emission tomography
Microwave receivers	Scans for breast cancer
Image-processing software (IRAF and AIPS) developed by NRAO, NOAO, and NASA	Cardiac angiography Monitoring neutron activity in brain
Positive pressure clean rooms for assembly of space instruments	Cleaner hospital operating rooms
Detection of faint x-ray sources	Portable x-ray scanners (Lixiscope) for neonatology and Third World clinics

in medical imaging that includes CAT scans, magnetic resonance imaging, and positron emission tomography. Microwave receivers developed by radio astronomers are used in scans for breast cancer.

Industry

Radio astronomy has been a particularly fruitful source of useful technology and algorithms. The Millitech Corporation, whose founders are radio astronomers, builds millimeter wavelength components, largely for the communications industry. The National Radio Astronomy Observatory (NRAO) has improved low-noise receivers, some of which have given rise to commercial products (Table 8.2). Computer programs used to control telescopes and to make maps from interferometers have found wide application in industry.

The effort to produce ever better emulsions for astronomical purposes led the Kodak Company to the discovery of gold sensitization, which made possible not only Tri-X film but an entire generation of 400-ASA films. Technical Pan, the film of choice for most industrial and fine arts photographers because of its fine grain and high resolution, first served to record changes on the surface of the sun. The infrared emulsions that astronomers first requested have proved useful in aerial reconnaissance and more recently in remote sensing of the earth's resources.

Defense Technology

The technical needs of some astronomical facilities and of certain national security programs are so similar that advances in one field often find applications

TABLE 8.2 Industrial Applications of Astronomical Techniques

Astronomical Technique or Device	Industrial Uses
Image-processing software (AIPS, IRAF)	General Motors Co. study of automobile crashes; Boeing Co. tests of aircraft hardware
Holographic methods for testing figures of radio telescopes	Testing communications antennas
Development of low-noise receivers by NRAO, universities	Components for communications industry
FORTH computer language developed by NRAO for control of radio telescopes	20 vendors supply FORTH for applications including analysis of auto engines in 20,000 garages, quality control for films at Kodak, 50,000 hand-held computers used by express mail firm
Gold sensitization of photographic plates	Development of Tri-X and 400-ASA films by Kodak
Infrared-sensitive films for spectroscopy	Aerial reconnaissance and earth resources mapping
X-ray detectors for NASA telescopes	Baggage scanners at airports
Gas chromatographs to search for life on Mars	

in the other. Progress in military technology, from World War II radar to present-day infrared detectors, has greatly enhanced astronomical capabilities. Yet astronomical data and techniques developed for astronomy have also proven useful for national security goals. Satellite and aerial surveillance requires lightweight telescopes, precise optical instruments, and the ability to process numerous imperfect images to extract the maximum available information. Development of the necessary mirror technology, the ability to adapt optics to rapidly changing conditions, and the processing algorithms have, from the U2 airplanes of the 1960s to the KH-11 satellites of the 1980s, involved people originally trained as astronomers.

Astronomers have also made important scientific and technical contributions to defense-related interests in the infrared wavelength range. For example, astronomers preparing instruments for the Hubble Space Telescope worked with researchers at the Honeywell Corporation and the Rockwell International Company to improve greatly the sensitivity of 1- to 2.5-μm infrared arrays. The Air Force Geophysics Laboratory rocket program produced pioneering data on the infrared sky, but NASA's Infrared Astronomical Satellite (IRAS) definitively

TABLE 8.3 Defense Applications of Astronomical Techniques

Astronomical Technique or Device	Defense Uses
Stellar observations and model atmospheres	Discrimination of celestial objects from rocket plumes, satellites, and warheads
Infrared all-sky survey by NASA's IRAS satellite	
Detectors for gamma-ray and x-ray astronomy	Vela satellite monitors for nuclear explosions Detection of nuclear reactors on Soviet spacecraft
Positions of quasars and stars	Precision navigation for civil and military purposes

measured the celestial infrared background against which orbiting satellites and incoming warheads must be detected (Table 8.3).

Astronomers working on x- and gamma-ray detectors at Los Alamos also helped build the instruments for the Vela satellites that monitored the earth for atomic explosions during the 1960s and 1970s. Two gamma-ray instruments operated for astronomical research independently confirmed the presence of nuclear reactors on several Soviet satellites.

Why They Call It Universal Time

Astronomical, civil, and defense interests overlap in their need for precise coordinate systems and timekeeping. Nonastrophysical uses include navigation, clock synchronization, ballistic missile guidance, and secure communications. The most accurate time standard for periods in excess of a few months may be extraterrestrial, with the recently discovered pulsars with millisecond periods proving to be the most accurate clocks in the galaxy.

The fundamental celestial coordinate system used for navigation is now based on radio astronomy. The locations of the satellites that make up the Global Positioning System are soon to be tied to the fixed positions on the sky of distant quasars. Inertial guidance systems (for missiles and other purposes) require this accurate astronomical coordinate system for their calibration. Finally, because satellite orbits are independent of the assorted wobbles of the earth beneath, the accurate location of terrestrial features requires accurate forecasts of the earth's orientation. The U.S. Naval Observatory disseminates this information based on optical and radio observations of quasars.

TABLE 8.4 Environmental Applications of Astronomical Techniques

Astronomical Technique or Device	Environmental Uses
Millimeter wave spectroscopy	Study of ozone depletion
Models of planetary atmospheres	Global change modeling
Measurement of variations of sun and solar-type stars	Study of global climatic change
Study of sunspots and solar flares in sun and stars	Short- and long-term prediction of terrestrial effects
Models of astrophysical shocks	Study of terrestrial storms
Precision measurement of quasars	Geodesy and study of tectonic drift
Composite materials for orbiting infrared telescope	Design of solar collectors
Theory of cosmic rays, solar flares, and stellar fusion	Design of fusion reactors

Energy

Key ideas in the controlled magnetic thermonuclear fusion program in the United States were provided by an astronomer who adapted ideas first considered in connection with cosmic rays and with nuclear fusion in stellar interiors (Table 8.4). The diagnostic tools developed for the study of solar flares and other hot, magnetized plasmas have proven useful in the investigation of the magnetic confinement of fusion plasmas on the earth. The search for alternative energy sources has also benefited from astronomical spin-offs. A private company has built solar radiation collectors up to 16 m in diameter using graphite composite materials first developed for a proposed orbiting telescope, the Large Deployable Reflector. The material is both light and resistant to temperature-induced distortions.

ASTRONOMY AS AN INTERNATIONAL ENTERPRISE

Astronomers have a long history of international collaboration, dating back to a network of comet observers established by Newton and Halley in the 17th century. The International Astronomical Union was the first of the modern international scientific unions organized under the Versailles treaty. The need for observatories all around the earth to cover the whole sky at all times encourages international collaborations. The names of the European Southern Observatory, the International Ultraviolet Explorer, and the Canada-France-Hawaii Telescope are self-explanatory. The latter shares the top of Mauna

Kea in Hawaii with a British infrared telescope and several American projects. Construction of a Japanese observatory on Mauna Kea is expected to begin in the next decade. Other success stories include NASA's IRAS satellite, a joint U.S., U.K., and Netherlands enterprise; an American spectrometer launched by a Japanese rocket to study cosmic radio waves; an American instrument on the Soviet Vega spacecraft that flew past Comet Halley; and a worldwide network of telescopes to study seismic oscillations of the sun. Scientists in the United States and the USSR will work together on a Soviet orbiting telescope for very long baseline interferometry called RadioAstron. This program builds on a history of collaboration between Soviet and American radio astronomers that survived the most difficult periods of the Cold War.

Civilization reaps the benefits of astronomy as an international enterprise. The first images from space showing the earth as a planet displayed the fragility of our planet and emphasized the need for worldwide cooperation in studying the earth and the universe it inhabits.

9

References

Abt, H. 1990. *Publ. Astron. Soc. Pacific* **102**, 1161.

American Astronomical Society (AAS). December 1990. *The American Astronomical Society 1990 Membership Survey.* AAS internal report. AAS, Washington, D.C.

Bracewell, R.A., and A.C. Riddle. 1967. *Astrophys. J.* **150**, 427.

Brown, R., ed. 1990. *An Educational Initiative in Astronomy.* Space Telescope Science Institute, Baltimore.

Burns, J., and W. Mendell, eds. 1988. *Future Astronomical Observatories on the Moon.* NASA CP2489. National Aeronautics and Space Administration, Washington, D.C.

de Zafra, R.L., M. Jaramillo, A. Parrish, P. Solomon, B. Connor, and J.W. Barrett. 1987. *Nature* **328**, 408.

Ellis, S.D., and P.J.M. Mulvey. 1989. *1987-88 Survey of Physics and Astronomy Bachelor's Degree Recipients.* Publ. No. R-211.20. American Institute of Physics, New York.

Hawking, S. 1988. *A Brief History of Time.* Bantam, New York.

Hickey, J.R., B.M. Alton, H.L. Kyle, and E.R. Major. 1988. *Adv. Space Res.* **8**, 5.

Mumma, M., and H. Smith, eds. 1990. *Astrophysics from the Moon.* American Institute of Physics Conference Proceedings No. 207. American Institute of Physics, New York.

Nakajima, T., S.R. Kulkarni, P.W. Gorham, A.M. Ghez, G. Neugebauer, J.B. Oke, T.A. Prince, and A.R. Readhead. 1989. *Astron. J.* **97**, 1.

National Academy of Sciences and National Academy of Engineering (NAS-NAE). 1988. *Toward a New Era in Space: Realigning Policies to New*

Realities (the "Stever Report"). Committee on Space Policy. National Academy Press, Washington, D.C.

National Aeronautics and Space Administration (NASA). 1988, 1989. *Strategic Plan.* Office of Space Science and Applications. NASA, Washington, D.C.

National Aeronautics and Space Administration (NASA). November 1989. *Report of the 90-Day Study on Human Exploration of the Moon and Mars.* NASA, Washington, D.C.

National Commission on Excellence in Education (NCEE). 1983. *A Nation at Risk: The Imperative for Educational Reform.* U.S. Department of Education.

National Research Council (NRC). 1964. *Ground-based Astronomy: A Ten-Year Program* (the "Whitford Report"). National Academy of Sciences, Washington, D.C.

National Research Council (NRC). 1972. *Astronomy and Astrophysics for the 1970's* (the "Greenstein Report"). National Academy of Sciences, Washington, D.C.

National Research Council (NRC). 1982. *Astronomy and Astrophysics for the 1980's. Volume I: Report of the Astronomy Survey Committee* (the "Field Report"). National Academy Press, Washington, D.C.

National Research Council (NRC). 1984. *A Strategy for the Explorer Program for Solar and Space Physics.* National Academy Press, Washington, D.C.

National Research Council (NRC). 1986a. *Physics Through the 1990s.* National Academy Press, Washington, D.C.

National Research Council (NRC). 1986b. *The Explorer Program for Astronomy and Astrophysics.* Committee on Space Astronomy and Astrophysics. National Academy Press, Washington, D.C.

National Research Council (NRC). 1988. *Space Science in the 21st Century.* Space Science Board. National Academy Press, Washington, D.C.

National Research Council (NRC). 1990a. *A Challenge of Numbers: People in the Mathematical Sciences.* National Academy Press, Washington, D.C.

National Research Council (NRC). 1990b. *Human Exploration of Space: A Review of NASA's 90-Day Study and Alternatives.* National Academy Press, Washington, D.C.

National Research Council (NRC). 1991. *Working Papers: Astronomy and Astrophysics Panel Reports.* National Academy Press, Washington, D.C.

Pan, X.P., M. Shao, M.M. Colavita, D. Mozurkewich, R.S. Simon, and K.J. Johnston. 1990. *Astrophys. J.* **356**, 641.

Porter, B.F., and D. Kellman. 1989. *AIP Membership Profile: Employment Mobility and Career Change.* American Institute of Physics Report R-306.2. American Institute of Physics, New York.

Sadler, P., K. Brecher, and O. Gingerich. 1989. Internal report. Center for Astrophysics, Cambridge, Mass.

Sagan, C. 1983. *Cosmos.* Random House, New York.

Solomon, P.M., R.L. de Zafra, A. Parrish, and J.W. Barrett. 1984. *Science* **224**, 1210.

APPENDICES

Appendix A

Glossary

ASTRONOMICAL TERMS

Accretion, accretion disk—Astronomical objects as diverse as *protostars* and *active galaxies* may derive their energy from the gravitational power released by the infall, or accretion, of material onto a central object. The combined effects of gravity and rotation often force the accreting material into an orbiting accretion disk.

Active galaxy—Certain galaxies emit far more energy than can be accounted for by their stars alone. The central regions of these galaxies harbor a compact, solar-system-sized object capable of outshining the rest of the galaxy by a factor of 100. The ultimate energy source for active galaxies may be the *accretion* of matter onto a *black hole*. Active galaxies can emit strongly across the entire *electromagnetic spectrum*, from radio waves to gamma rays. See *quasar*.

Active optics—A technique to take account of slowly varying forces, such as gravitational deflections and temperature drifts, that can distort a mirror on time scales of minutes to hours, resulting in imperfect images.

Adaptive optics—A set of techniques to adjust the mirrors of telescopes on time scales of hundredths of a second to correct for distortions in astronomical images due to turbulence in the earth's atmosphere.

Airshower detector—Highly energetic gamma rays collide with atoms and molecules in the earth's upper atmosphere and produce bursts of light and particles, called airshowers, that can be detected from the ground, giving information about the most energetic processes in the universe.

Anisotropy—The distribution of galaxies in space is not uniform (isotropic) to the limit of the most sensitive surveys. However, the intensity of the *cosmic background radiation* from the *Big Bang* is highly uniform in all directions. Astronomers are searching with more sensitive telescopes for the small anisotropies in the cosmic background radiation that should be present given the non-uniform distribution of galaxies.

Aperture synthesis—see *interferometer*.

Arcminute—A unit of angle corresponding to 1/60th of a degree. The full moon is 30 arcminutes in diameter.

Arcsecond—A unit of angle corresponding to 1/3600th of a degree; 1/60th of an *arcminute*. An arcsecond is approximately the size of a penny viewed from about 2.5 miles.

Array—(1) Groups, or arrays, of telescopes can be combined to simulate a single large telescope, kilometers or even thousands of kilometers across. (2) Astronomical instruments have recently been fabricated using new electronic components called detector arrays or *charge-coupled devices* (CCDs) that consist of thousands of individual detectors constructed on centimeter-sized wafers of silicon, or other materials.

Astrometry—The branch of astronomy concerned with measuring the positions of celestial objects. Advances in technology may soon permit a 1,000-fold improvement in the measurement of positions, and thus in our ability to determine distances to stars and galaxies. See *parallax*.

Baseline—The separation between telescopes in an *interferometer*. The largest baseline determines the finest detail that can be discerned with an interferometer.

Big Bang—Most astronomers believe that the universe began in a giant explosion called the "Big Bang" some 10 billion to 20 billion years ago. Starting from an initial state of extremely high density, the universe has been expanding and cooling ever since. Some of the most fundamental observed properties of the universe, including the abundance of light elements such as helium and lithium and the recession of galaxies, can be accounted for by modern theories of the Big Bang.

Black hole—A region in space where the density of matter is so extreme, and the resultant pull of gravity so strong, that not even light can escape. Black holes may be the endpoint in the evolution of some types of stars and may be located at the centers of some *active galaxies* and *quasars*.

Blackbody radiation—A glowing object emits radiation in a quantity and at wavelengths that depend on the temperature of the object. For example, a poker placed in a hot fire first glows red-hot, then yellow-hot, then finally white-hot. This radiation is called thermal or blackbody radiation.

Brown dwarf—A star-like object that contains less than about 0.08 the mass of the sun and is thus too small to ignite nuclear fuels and become a normal star. Brown dwarfs emit small amounts of infrared radiation due to the slow release of gravitational energy and may be a component of the *missing mass*.

Byte—A unit of information used in reference to computers and quantities of data. A byte consists of 8 bits (0's and 1's) and may correspond to a single character or number.

Carbon monoxide—A molecule (CO) consisting of carbon and oxygen that emits strongly at millimeter and *submillimeter* wavelengths and that can be used to trace cool gas in our own and other *galaxies*.

Charon—A recently discovered moon of the planet Pluto.

Charge-coupled device, or CCD—An electronic detector used for low-light-level imaging and astronomical observations. CCDs were developed by NASA for use in the Hubble Space Telescope and the Galileo probe to Jupiter and are now widely used on ground-based telescopes.

Cosmic background radiation—The radiation left over from the *Big Bang* explosion at the beginning of the universe. As the universe expanded, the temperature of the fireball cooled to its present level of 3 degrees above absolute zero (−270 degrees celsius). *Blackbody radiation* from the cosmic background is observed at radio, millimeter, and *submillimeter* wavelengths.

Cosmic rays—Protons and nuclei of heavy atoms that are accelerated to high energies in the magnetic field of our galaxy and that can be studied directly from the earth or from satellites.

Dark matter—Approximately 90 percent of the matter in the universe may so far have escaped direct detection. The presence of this unseen matter has been inferred from motions of stars and gas in *galaxies*, and of galaxies in clusters of galaxies. Candidates for the missing mass include *brown dwarf* stars and exotic subatomic particles. Dark matter was called "missing mass" for many years. However, because it is the light, not the mass, that is missing, astronomers have given up this terminology. Also called *missing mass*.

Diffraction limit—The finest detail that can be discerned with a telescope. The physical principle of diffraction limits this to a value proportional to the wavelength of the light observed divided by the diameter of the telescope.

Einstein Observatory—An x-ray telescope launched in 1978, one of a series of High-Energy Astrophysics Observatories, HEAO-2.

Electromagnetic spectrum—Radiation can be represented as electric and magnetic fields vibrating with a characteristic wavelength or frequency. Long wavelengths (low frequencies) correspond to radio radiation, intermediate wavelengths to millimeter and infrared radiation, short wavelengths (high frequencies) to visible and ultraviolet light, and extremely short wavelengths to x-rays and gamma rays. Most astronomical observations measure some form of electromagnetic radiation.

Expansion of the universe—The tendency of every part of the universe to move away from every other part due to the initial impetus of the *Big Bang;* also known as the Hubble expansion, after the American astronomer Edwin Hubble, whose observations of receding *galaxies* led to our present understanding of the expanding universe. See *redshift.*

Extragalactic—Objects outside our *galaxy*, more than about 50,000 light-years away, are referred to as extragalactic.

Field Report—A National Research Council report (NRC, 1982) on astronomy in the 1980s chaired by G.B. Field.

Fly's Eye—A telescope used to monitor gamma rays from astronomical sources.

Galaxy—An isolated grouping of tens to hundreds of billions of stars ranging in size from 5,000 to 20,000 light-years across. Spiral galaxies like our own *Milky Way* are flattened disks of stars and often contain large amounts of gas out of which new stars can form. Elliptical galaxies are shaped more like footballs and are usually devoid of significant quantities of gas.

Gamma-ray astronomy—The study of astronomical objects using the most energetic form of electromagnetic radiation.

Gigabyte—One billion (10^9) *bytes*. A unit of information used to describe quantities of data or the storage capacity of computers.

Gravitational lens—A consequence of Einstein's general relativity theory is that the path of light rays can be bent by the presence of matter. Astronomers have observed that the light from a distant galaxy or quasar can be "lensed" by the matter in an intervening galaxy to form multiple and often distorted images of the background object.

Great Observatories—A NASA program to launch four major observatories to cover the optical (HST), gamma-ray (GRO), X-ray (AXAF), and infrared (SIRTF) portions of the electromagnetic spectrum.

Green Bank Telescope—A 100-m steerable radio telescope under construction in Green Bank, W. Va.

Greenstein Report—A National Research Council report (NRC, 1972) on astronomy in the 1970s chaired by J.L. Greenstein.

Helioseismology—The study of the internal vibrations of the sun. In a manner analogous to terrestrial seismology, helioseismology can reveal important information about the sun's internal condition.

Hubble Space Telescope (HST)—A 2.4-m-diameter telescope orbiting in space, designed to study visible, ultraviolet, and infrared radiation; the first of NASA's Great Observatories.

Hydrogen—The most abundant element in the universe. It can be observed at a variety of wavelengths, including 21-cm radio, infrared, visible, and ultraviolet wavelengths, and in a variety of forms, including atoms (HI), molecular form (H_2), and ionized form (HII).

Infrared astronomy—The study of astronomical objects using intermediate-wavelength radiation to which the atmosphere is mostly opaque and the human eye insensitive. Humans sense infrared energy as heat. The infrared part of the *electromagnetic spectrum* generally corresponds to radiation with wavelengths from 1 μm to 1000 μm (1 μm is one-millionth of a meter). Objects with temperatures around room temperature or lower emit most of their radiation in the infrared.

Interferometer, interferometry—A spatial interferometer combines beams of light from small, widely separated telescopes to synthesize the aperture of a single large telescope; see *array*. A different form of interferometer can be used on a single telescope to break up the light into its constituent colors; see *spectroscopy*.

Io—A moon of the planet Jupiter on which volcanoes have been observed from the Voyager spacecraft and from terrestrial telescopes.

Isoplanatic angle—The largest field of view over which a distortion-free image can be formed looking through the earth's atmosphere; a few *arcseconds* at visible wavelengths.

Kamiokande—A Japanese underground observatory used to detect *neutrinos*.

Kuiper Airborne Observatory—A 1-m-diameter telescope for *infrared* and *submillimeter* observations that is carried above most of the earth's water vapor in a C-141 aircraft.

Light-year—A unit of astronomical distance equal to the distance light travels in a year: about 9 trillion miles. The nearest star is 4 light-years away. The center of our galaxy is about 24,000 light-years away. The closest galaxy is about 180,000 light-years away.

Magellanic Clouds, Large and Small—The two closest galaxies to our own Milky Way, located about 180,000 light-years away and visible only from the Southern Hemisphere. A bright *supernova*, SN1987A, was observed in the Large Magellanic Cloud in 1987.

Magnetohydrodynamics—The study of the motion of gases in the presence of magnetic fields.

Magnitude—A unit of brightness for stars. Fainter stars have numerically larger magnitudes. The brightest stars, excluding the sun, are about magnitude 0; the faintest star visible to the unaided eye is about magnitude 6. A star with V = 15 is one-millionth as bright as the half-dozen brightest stars with V = 0. Stars as faint as magnitude 28 can be seen with powerful terrestrial or spaceborne telescopes.

Megabyte—One million *bytes*. A unit of information used to describe quantities of data or the storage capacity of computers. A single image from the Hubble Space Telescope comprises about 5 megabytes.

Milky Way—Our sun is located in the Milky Way Galaxy, which consists of some 100 billion stars spread in a disk over 50,000 light-years across and hundreds of light-years thick. (See the cover of this report.)

Missing mass—Approximately 90 percent of the matter in the universe may so far have escaped direct detection. The presence of this unseen matter has been inferred from motions of stars and gas in *galaxies*, and of galaxies in clusters of galaxies. Candidates for the missing mass include *brown dwarf* stars and exotic subatomic particles. Dark matter was called "missing mass" for many years. However, because it is the light, not the mass, that is missing, astronomers have given up this terminology. Also called *dark matter*.

Neutrino—One of a family of subatomic particles with little or no mass. These particles are generated in nuclear reactions on the earth, in the centers of stars, and during *supernova* explosions and can give unique information about these energetic processes. Because neutrinos interact only weakly with matter, they are difficult to detect.

Nucleosynthesis—The process by which heavy elements such as helium, carbon, nitrogen, and iron are formed out of the fusion of lighter elements, such as hydrogen, during the normal evolution of stars, during *supernova* explosions, and in the *Big Bang*.

Optical astronomy—The study of astronomical objects using light waves with wavelengths from about 1 to 0.3 μm. The human eye is sensitive to most of these wavelengths. See *electromagnetic spectrum*.

Parallax—The apparent shift in position of a nearby object relative to a more distant object, as the observer changes position. Using basic trigonometry, it is possible to derive the distance of a star from its parallax as observed from opposite points on the earth's orbit. See *astrometry*.

Parsec—A unit of astronomical distance equal to 3.3 light-years.

Pixel—The smallest element of a digital image. A typical image from the *Hubble Space Telescope* is a square with 1,600 $\times$ 1,600 discrete pixels.

Planetary debris disk—A cloud of solid material (dust) orbiting a star; possibly the solid material left over from the formation of a planetary system.

Protogalaxy—*Galaxies* are thought to have formed fairly early in the history of the universe, by the collapse of giant clouds of gas. During this process, a first generation of stars formed, and these should be observable with the telescopes discussed in this report.

Protoplanetary or protostellar disk—A disk of gas and dust surrounding a young star or *protostar* out of which planets may form.

Protostar—The earliest phase in the evolution of a star, in which most of its energy comes from the infall of material, or *accretion*, onto the growing star. A *protostellar disk* probably forms around the star at this time.

Quasar—An extremely compact, luminous source of energy found in the cores of certain galaxies. A quasar may outshine its parent galaxy by a factor of 1,000 yet be no larger than our own solar system. The *accretion* of gas onto a *black hole* may power the quasar. *Active galaxies* are probably less luminous, more-nearby versions of quasars.

Radio astronomy—The study of astronomical objects using radio waves with wavelengths generally longer than 0.5 to 1 mm. See *electromagnetic spectrum*.

Redshift—Radiation from an approaching object is shifted to higher frequencies (to the blue), while radiation from a receding object is shifted to lower frequencies (to the red). A similar effect raises the pitch of an ambulance siren as it approaches. The *expansion of the universe* makes objects recede so that the light from distant galaxies is redshifted. The redshift is often denoted by z, where $z = v/c$ and v is the velocity and c the speed of light. The wavelength shift is then given by the factor $(1 + z)$.

Regolith—Lunar topsoil.

Resolution—Spatial resolution describes the ability of an instrument to separate features at small details; see *diffraction limit* and *interferometer*. Spectral resolution describes the ability of an instrument to discern small shifts in wavelength; see *spectroscopy*.

Roentgen Satellite—See *ROSAT*.

Space Exploration Initiative—A plan proposed by the President for the manned exploration of the moon and Mars.

Spectroscopy—A technique whereby the light from astronomical objects is broken up into its constituent colors. Radiation from the different chemical elements that make up the object can be distinguished, giving information about the abundances of these elements and their physical state.

Star-burst galaxy—Certain *galaxies*, particularly those perturbed by a close encounter or collision with another galaxy, often form stars at a rate hundreds of times greater than in our galaxy. Such galaxies are bright sources of *infrared* radiation.

Submillimeter radiation—*Electromagnetic* radiation with wavelengths between about 0.1 and 1 mm intermediate between radio and infrared radiation.

Sunyaev-Zeldovich effect—An astrophysical effect whereby the distribution of wavelengths of radiation seen through the gas in a distant cluster of galaxies is subtly modified. Measurement of this effect can be used to determine the distance to the cluster.

Supercomputer—At any one time, the fastest, most powerful computer available at any price.

Supernova—A star that, due to *accretion* of matter from a companion star or exhaustion of its own fuel supply, can no longer support itself against its own weight and collapses, throwing off its outer layers in a burst of energy that outshines an entire galaxy. In 1987 a star in the *Large Magellanic Cloud* was observed as a dramatic supernova called Supernova 1987A.

Supernova 1987A—See *supernova*.

Terabyte—One trillion (10^{12}) *bytes*. A unit of information used to measure quantities of data. All the images taken with the Hubble Space Telescope in a given year will comprise a few terabytes.

Ultraviolet (UV) astronomy—The study of astronomical objects using short-wavelength radiation, from 0.3 μ to 0.01 μm (10 nm), to which the atmosphere is opaque and the human eye insensitive. See *electromagnetic spectrum.*

V, visual magnitude—See *magnitude.*

Whitford Report—A National Research Council report (NRC, 1964) on astronomy in the 1960s chaired by A.E. Whitford.

Working Papers—The unrefereed reports (NRC, 1991) of 14 of the 15 panels constituted as part of this survey of astronomy and astrophysics.

X-ray astronomy—The study of astronomical objects using x-rays having wavelengths shorter than about 10 nm to which the atmosphere is opaque. X-rays are emitted by extremely energetic objects having temperatures of millions of degrees. See *electromagnetic spectrum.*

X-ray background—The sky as observed at x-ray wavelengths is not completely dark, but glows faintly. The origin of this diffuse emission remains puzzling: possibly the radiation from countless galaxies, possibly the radiation from gas filling the space between galaxies. AXAF and other space missions may discover the secret in the 1990s.

z—See *redshift.*

ABBREVIATIONS AND ACRONYMS

ACE—Advanced Composition Explorer. A space mission to study cosmic rays.

AIM—The proposed Astrometric Interferometry Mission (AIM) would use small telescopes in space separated by up to 100 m to measure the positions of stars with 3- to 30-millionths-of-an-arcsecond precision.

AIPS—Astronomical Image Processing System. A set of programs developed to process astronomical data from the Very Large Array (VLA) and other radio wavelength *interferometers.*

ASTRO—A collection of telescopes that flew on the Space Shuttle in December 1990 to measure ultraviolet and x-ray radiation.

ASTRO-D—A Japanese-built x-ray telescope to be flown in 1993 in collaboration with U.S. astronomers.

AU—Astronomical Unit. A basic unit of distance equal to the separation between the earth and the sun, about 150 million km.

AXAF—The Advanced X-ray Astrophysics Facility. A telescope, now under construction, designed to observe x-rays. The third of NASA's Great Observatories.

CASA—Chicago Airshower Array for the study of gamma rays.

CCD—See *charge-coupled device.*

CO—See *carbon monoxide.*

COBE—The Cosmic Background Explorer. A NASA mission launched in 1989 to study the *cosmic background radiation* from the *Big Bang.*

CSO—The Caltech Submillimeter Observatory, a 10-m telescope operating on Mauna Kea, Hawaii. The telescope is used for observations of millimeter and *submillimeter* wavelength radiation.

DOD—Department of Defense.

DOE—Department of Energy.

ESA—The European Space Agency. The European equivalent of NASA.

ESO—The European Southern Observatory.

EUVE—The Extreme Ultraviolet Explorer. A NASA mission planned for the 1990s.

FITS—Flexible Image Transport System. A worldwide format for transferring astronomical images and other information between computers.

FUSE—The Far Ultraviolet Spectroscopy Explorer. A NASA mission planned for the 1990s.

GBT—The Green Bank Telescope.

GMRT—The Giant Meter Wave Radio Telescope in India.

GONG—Global Oscillations Network Group. A worldwide network of telescopes designed to study vibrations in the sun. See *helioseismology*.

GRO—The Gamma Ray Observatory. A telescope to be launched in 1991 to study highly energetic gamma rays from astronomical sources. NASA's second Great Observatory.

HEAO—The High-Energy Astronomical Observatory. A series of three telescopes launched in the 1980s to study x-rays and gamma rays. HEAO-2 was also called the *Einstein Observatory.*

HETE—The High-Energy Transient Experiment. An experiment to be launched from the Space Shuttle to look for the origins of mysterious bursts of x-rays and gamma rays.

HI, HII, H_2—See *hydrogen.*

HST—*Hubble Space Telescope.*

IRAF—Image Reduction and Analysis Facility. A set of computer programs for working with astronomical images.

IRAS—The Infrared Astronomical Satellite. A NASA Explorer satellite launched in 1983 that surveyed the entire sky in four infrared wavelength bands using a helium-cooled telescope.

IRTF—The Infrared Telescope Facility. A 3-m telescope located on Mauna Kea, Hawaii, and operated by NASA to study planets and other astronomical objects.

ISO—The Infrared Space Observatory. A European mission planned for launch in 1994 to study infrared radiation.

IUE—The International Ultraviolet Explorer. A joint NASA-ESA orbiting telescope to study ultraviolet radiation.

KAO—See *Kuiper Airborne Observatory.*

LEST—The Large Earth-based Solar Telescope.

MMA—The Millimeter Array. A proposed instrument that would link 40 telescopes together as an *interferometer* to study millimeter wavelength radiation.

MPIR—Max-Planck-Institute for Radio Astronomy.

NAIC—National Astronomy and Ionosphere Center.

NAS—National Academy of Sciences.

NASA—National Aeronautics and Space Administration.

NICMOS—The Near-Infrared Camera and Multi-Objective Spectrometer. A second-generation instrument for infrared imaging and spectroscopy planned for the *Hubble Space Telescope*.

NOAO—National Optical Astronomy Observatories.

NRAO—National Radio Astronomy Observatory.

NRC—National Research Council.

NSF—National Science Foundation.

NSO—National Solar Observatory.

OSSA—NASA's Office of Space Science and Applications.

RadioAstron—A proposed Soviet mission to fly a radio telescope in space as part of an earth-space *interferometer*.

ROSAT—The Roentgen Satellite, an orbiting x-ray telescope launched in 1990, is named after the German scientist W. Röntgen, the discoverer of x-rays. ROSAT is a German-U.S.-U.K. collaboration.

SETI—Search for extraterrestrial intelligence.

SIRTF—The Space Infrared Telescope Facility. NASA's fourth Great Observatory will study infrared radiation.

SMEX—Small Explorer. A NASA program to fly small, inexpensive satellites on a rapid time scale.

SOFIA—The Stratospheric Observatory for Far-Infrared Astronomy. A 2.5-m telescope flown above most of the earth's water vapor in a modified 747 aircraft to study *infrared* and *submillimeter* radiation.

SOHO—The Solar-Heliospheric Observatory, a European Space Agency mission that will be launched around 1995 to study the sun.

STIS—The Space Telescope Imaging Spectrometer. A second-generation instrument for ultraviolet imaging and spectroscopy planned for the *Hubble Space Telescope*.

SWAS—The Submillimeter Wave Astronomy Satellite. A Small Explorer (*SMEX*) payload to study *submillimeter* emission from water and oxygen.

VLA—The Very Large Array. A radio interferometer consisting of 27 antennae spread out over 35 km and operating with 0.1-*arcsecond* spatial *resolution*.

VLBA—The Very Long Baseline Array. An array of radio telescopes operating as an interferometer with a transcontinental *baseline* and spatial *resolution* less than a thousandth of an *arcsecond*.

VLBI—Very long baseline interferometry. A technique whereby a network of radio telescopes can operate as an *interferometer* with *baselines* that can be as large as the entire earth, or even larger when satellites are used.

VLT—The Very Large Telescope. A European project to build four 8-m telescopes.

VSOP—A Japanese mission to fly a radio telescope in space to operate as part of an earth-space *interferometer*.

WF/PC—The Wide-Field/Planetary Camera. The main imaging instrument on the *Hubble Space Telescope*. A modified version will be installed in 1993 to correct for distortions in the *HST* mirror.

WIYN—University of Wisconsin, Indiana University, Yale University, and NOAO 3.5-m telescope.

XTE—The X-ray Timing Explorer. A NASA mission to study x-ray radiation.

Appendix B

Status of the Profession

This appendix describes the demographics of astronomy, including the support by the two major funding agencies, NSF and NASA, and lists major ground-based facilities in operation. These data provide a context for the recommendations made in the main text of the report. Related topics are discussed in Chapter 3 and in the reports of the panels on Policy Opportunities and on the Status of the Profession in the *Working Papers* (NRC, 1991).

THE DEMOGRAPHICS OF ASTRONOMY

The Growth of Astronomy

Statistics from the American Astronomical Society (1990) and the American Institute of Physics (AIP; Porter and Kellman, 1989) characterize the demographics of the profession and show that the astronomical community has grown steadily over the last decade. After correcting the current membership of the American Astronomical Society (AAS), over 5,100, for foreign members and other small factors, the Status of the Profession Panel (*Working Papers*) estimated that the United States has approximately 4,200 working astronomers. This value represents a 42 percent increase since 1980 (Figure B.1). Publication records analyzed by Abt (1990) show that approximately 2,800 astronomers, about two-thirds of the profession, are actively engaged in research; the AAS and AIP surveys show that the remainder are involved mostly in teaching or administration.

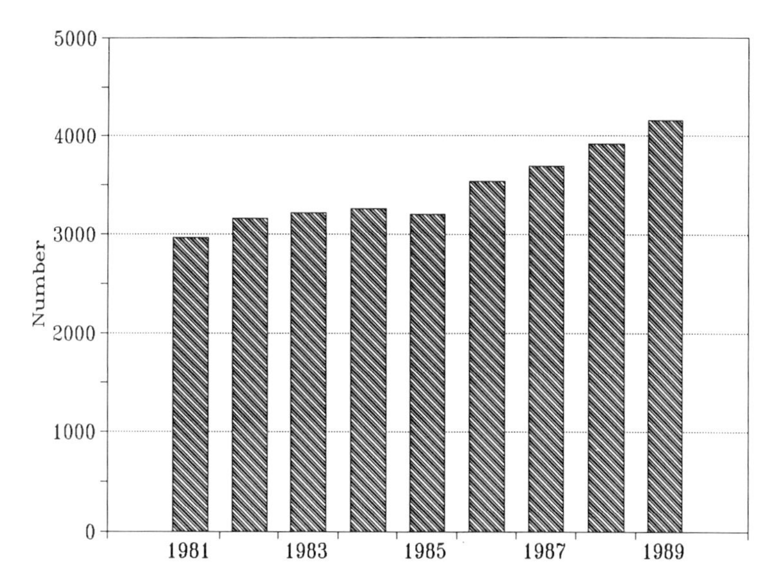

FIGURE B.1 The growth in the number of U.S. astronomers can be inferred from the membership of the American Astronomical Society (AAS), which has increased by 42 percent in the last decade. The numbers plotted were obtained by multiplying the AAS membership by 0.82 to account for foreign members in the AAS and other small corrections.

The number of doctoral degrees awarded for dissertations in astronomical topics over the last decade has averaged 125 degrees per year. However, the great scientific interest of astronomy has resulted in a flow of new people into the AAS, approximately 250 per year, that is about twice the rate of the production of PhDs in astronomy and astrophysics. Roughly half the people entering astronomy in the past decade were trained in a different science and subsequently migrated into the field.

Astronomy is being carried out by a young community. In 1987, the median age of U.S. astronomers was 42, the youngest of the 10 professional societies sampled by the AIP. Half the U.S. astronomers are currently between the ages of 35 and 50.

The percentage of women in astronomy has grown by almost 50 percent during the last decade, from 8 percent to 12 percent. Examination of the graduate student population shows signs of a slow, continuing growth in the number of women in astronomy. Recent surveys show that 12 percent of the doctoral degrees awarded by astronomy departments in 1987 went to women and that 20 percent of the total astronomy graduate population is female. These fractions are similar to or higher than the female membership of the AAS and suggest that the percentage of women in astronomy will continue to rise slowly during the 1990s.

Despite active recruitment efforts at universities and by NASA, ethnic minorities remain significantly underrepresented in astronomy. Preliminary results from a survey of the AAS membership show that 93 percent of the AAS members classify themselves as white, 4 percent as Asian, and 1 percent as Hispanic. Afro-Americans constitute less than 0.5 percent of the AAS membership. These statistics reflect the situation in other sciences and must be addressed by an increased emphasis on science education and recruitment at all levels—primary and secondary school and college—as discussed in Chapter 7.

Astronomy as a Profession

The types of jobs that astronomers hold have changed over the last decade. The rapid growth of astronomy in the late 1960s and the concomitant youthful age of the astronomical faculty has meant that faculty retirements have been few and far between. This has, in turn, severely limited the number of tenured faculty positions that opened up during the last 10 years. Industrial and national research facilities employ increasing numbers of astronomers to carry out important astronomical research.

Academic institutions currently employ roughly half of the AAS members. Nearly 75 percent of those employed in academia hold permanent, tenured positions funded by state or private resources. Not including astronomers holding postdoctoral appointments, 17 percent of the academic astronomers hold jobs as research associates, semipermanent academic positions dependent on outside funding.

In the 1980s, the fraction of PhD astronomers working at universities decreased markedly, while the number of researchers depending on federal funding rose [see the report of the Status of the Profession Panel, *Working Papers* (NRC, 1991)]. A decade ago, 40 percent of the class of 1970 was working at universities, and 30 percent in tenured or tenure-track positions. In 1989, only 32 percent of the class of 1980 was working at universities, with 22 percent holding tenured or tenure-track positions. Twelve percent of the class of 1980 was working for industry, compared with 1 percent for the class of 1970. Fifteen percent of the class of 1980 was working in national laboratories, compared with 11 percent of the class of 1970 a decade ago. Finally, the rate of people leaving the field 10 years after receiving a PhD is constant at about 30 percent.

Industrial jobs, although often dependent on a company's ability to obtain outside funding, also have a degree of permanence. The percentage of U.S. astronomers who work in a corporate setting has nearly doubled over the last decade. In 1987, the number of astronomers who worked for industry or were self-employed exceeded the number who worked in Federally Funded Research and Development Centers, such as the national observatories.

TABLE B.1 Estimated Nonfederal Salary Support for U.S. Astronomers

U.S. Astronomers	Salary Support (1990 $M)
Approximate total—4,200	
Median salary (1990$)—57,420	
Average overhead rate—1.50	
Fraction tenured or in tenure-track positions—0.51[a]	
Fraction of year supported (academics)—0.84	
Total academic support	154
Fraction employed in industry—0.09	
Fraction of year supported (industry)—0.50	
Total industry support	16
TOTAL NONFEDERAL SUPPORT	170

[a] Porter and Kellman (1989).

THE FUNDING OF ASTRONOMICAL RESEARCH

Federal funding through grants and through the national observatories supports the research of most of the astronomical community. The NSF provides the major share of funding for ground-based astronomy, while NASA provides all the funding for space astronomy. Historically, ground-based optical astronomy has attracted additional private and state support, radio astronomy has depended heavily on federal funding, and space astronomy has been entirely federally supported. The report of the Status of the Profession Panel (*Working Papers*) notes that of all the articles appearing in the *Astrophysical Journal* in 1989, 40 percent cited NASA support, 40 percent listed NSF support, and 10 percent acknowledged other sources of federal funding. The NSF and NASA grant programs are of paramount importance to the health of the field. Important support for special projects is provided by the DOE, the DOD, and the Smithsonian Institution.

Astronomy has a long tradition of private and state support, including university salaries, modest support for faculty research, and occasional philanthropic donations of large research instruments like the Hale 5-m telescope and the Keck 10-m telescope. Using the data from Porter and Kellman (1989), it is possible to estimate that the nonfederal funding of astronomy in terms of salaries amounts to about $170 million per year (Table B.1). Including support for construction and operation of private or state facilities raises the total nonfederal level of support for astronomy to around $190 million per year. It should be pointed out that much of the salary support for university astronomers covers their teaching, rather than their research, activities.

The following analysis of funding trends is confined to the NSF and NASA

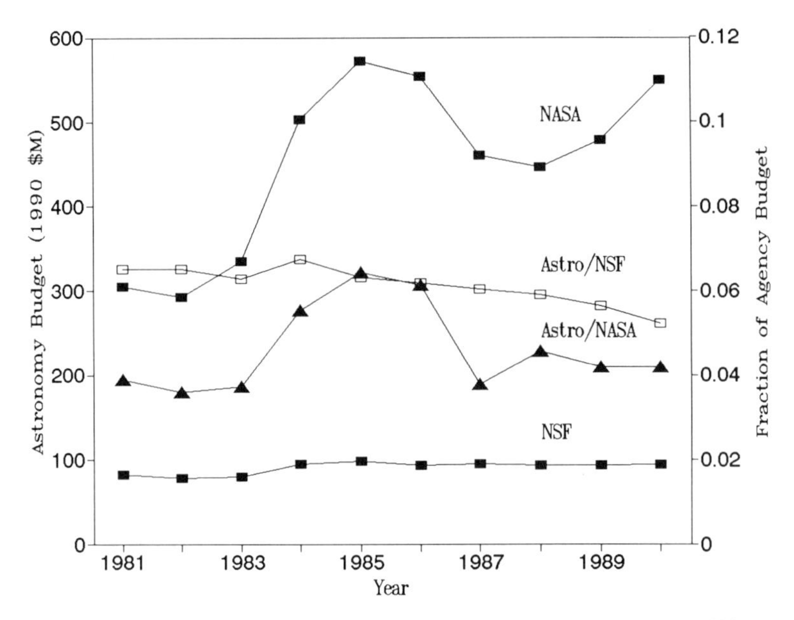

FIGURE B.2 Overall funding of astronomy by NASA and the NSF in constant (1990) dollars. The scale on the right shows the fractions of the two agencies' budgets devoted to astronomical activities.

astronomy programs. Figure B.2 shows the overall astronomy budgets for the two agencies and the fraction of the agency budgets devoted to astronomy. All references to funds have been converted to 1990 dollars using the Consumer Price Index cost-of-living index.

Support from the National Science Foundation

Historically, astronomy has received between 5 and 6 percent of the overall NSF budget. However, there has been a steady decline in this level to below 5 percent over the past decade (see Figure B.2). Funding from the NSF supports three major areas, the national centers, university observatories, and individual research grants (Figure B.3).

NSF grants range from small sums for the publication of conference proceedings to over $1 million per year for the operation of university observatories. The bulk of the grants program falls between these extremes and supports individual research projects. These grants make possible the research of individuals throughout the United States and provide for the training of the next generation of astronomers. Although the number of grants increased over the decade, Figure B.4 shows a steady erosion in the size of grants to individuals (excluding major grants for university observatories). With typical costs

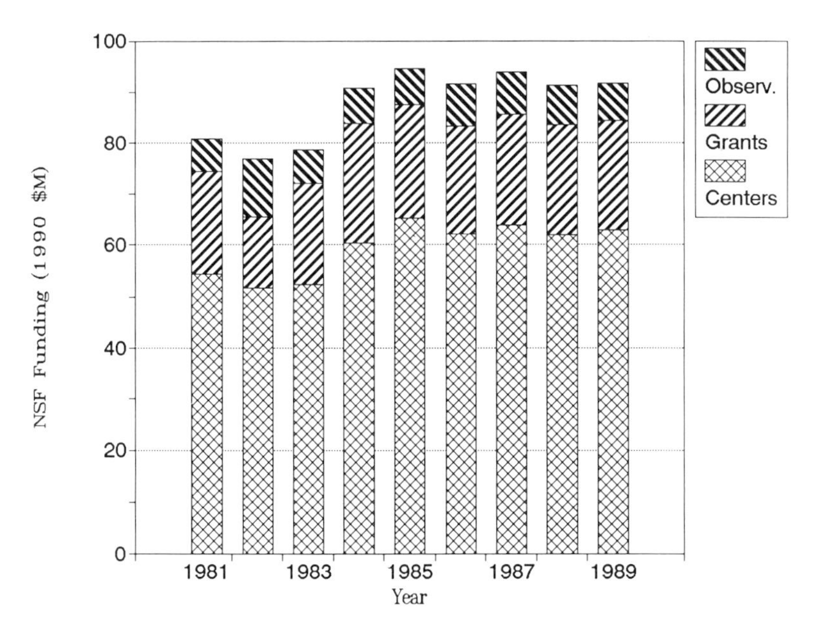

FIGURE B.3 The division between NSF funding of research grants and for the operation and construction of university observatories (observ.) and the national observatories (centers) over the last decade.

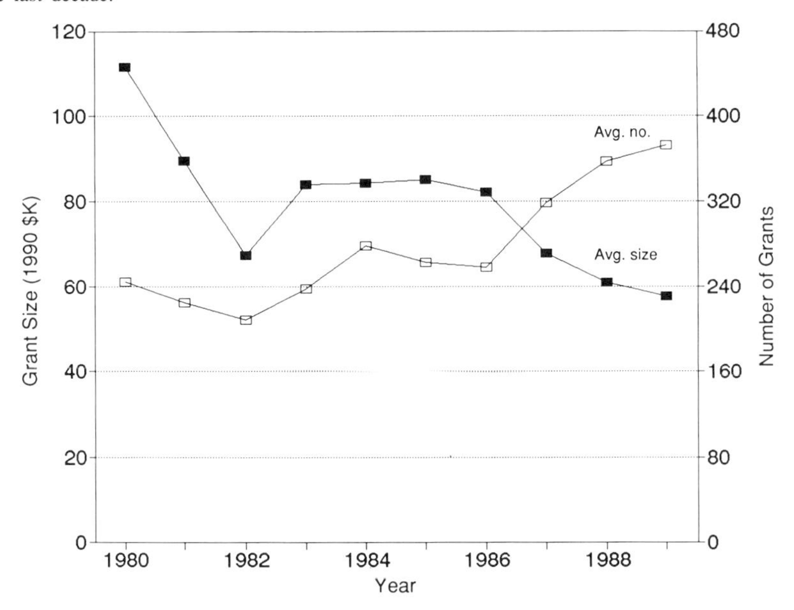

FIGURE B.4 The average size (1990 $K—solid rectangles) and number (open rectangles) of NSF research grants awarded in astronomy and astrophysics, corrected for the portion of the research budget devoted to operation and construction of university facilities.

of a summer salary, publications, and a graduate student salary, plus overhead, now exceeding $70,000 per year, the NSF grants program has dropped below a critical threshold. While an individual can, and often must, have more than one grant, such a system of support suffers from inefficiencies and often produces more proposals than research.

Support from NASA

Astronomy accounts for about 4 percent of the overall NASA budget and for about 25 percent of the budget of the Office of Space Science and Applications (see Figure B.2). Most of this money is spent on capital expenses associated with major observatory programs and increased temporarily during the mid-1980s, due in part to delays in the construction and launch of the Hubble Space Telescope. Figure B.5 shows the division between large programs (the Great Observatories); the moderate-scale Explorer program; and small projects, including the grants program, data analysis, and the Kuiper Airborne Observatory.

The NASA grants program increased during the 1980s, tracking the growth in the number of astronomers (Figure B.6). From 1982 to 1989, the average grant size remained stable at about $50,000. The small average size is due to the large number of small grants given to pay for observing with the International Ultraviolet Explorer (IUE). It should be pointed out, however, that the average grant is too small to support a postdoctoral fellow or, in some cases, even a graduate student. Many IUE researchers write multiple proposals to support themselves, their postdoctoral fellows, and their graduate students. In the rest of the NASA's astrophysics program, the grant sizes are significantly larger, about $83,000, although their size has decreased over the decade. The committee supports the efforts of NASA's Astrophysics Division to consolidate its large number of disparate grants programs into a smaller number.

NASA is becoming the dominant agency in astronomy grant funding. In 1982, NSF provided about 60 percent of the federal support for individual grants. By the end of the decade, NASA had provided more money for astronomy grants than had the NSF. Since several large, long-lived missions will be launched in the 1990s, NASA's grant support for data analysis is expected to increase even more.

ACCESS TO GROUND-BASED TELESCOPES

The telescopes currently used by the majority of astronomers are ground-based instruments operating at radio, optical, and infrared wavelengths. This section lists briefly the major instruments in use around the world at these wavelengths.

Americans want to participate directly in the thrill of discovery both

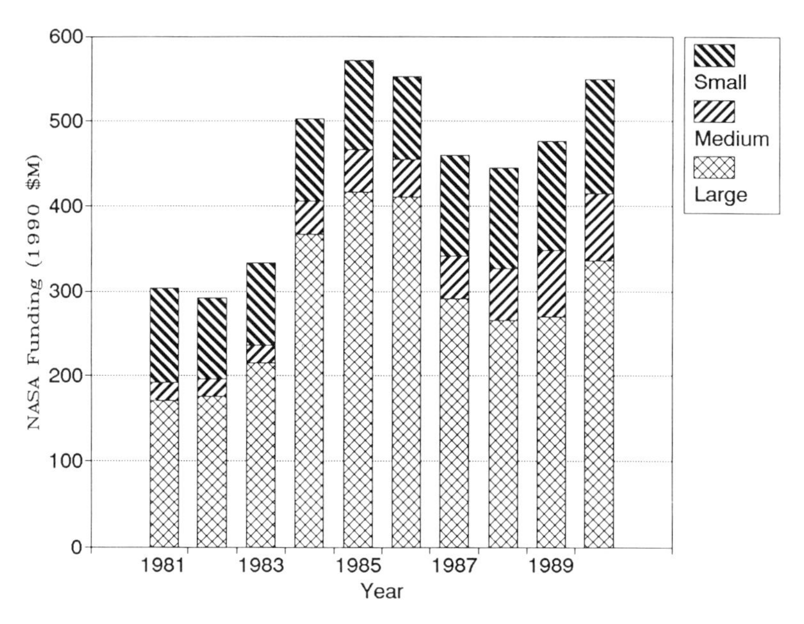

FIGURE B.5 NASA astronomy budget (1990 $M) divided between large programs, including the Great Observatories; moderate programs like the Explorers; and small projects, including research grants and the airborne program.

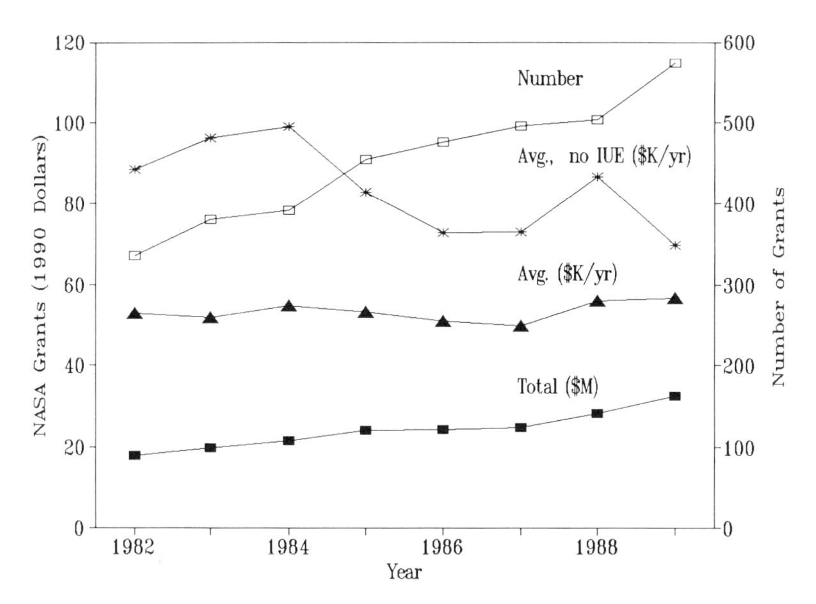

FIGURE B.6 NASA astronomy research grants program. The total program is shown in millions of dollars. Also shown are the average grant size (1990 $K/yr) with and without the IUE, and the total number of grants.

because of the excitement and the enlightenment that new knowledge of the universe gives and also because of the inspirational effect on young people who are considering careers in engineering and science. If we were not to have competitive ground-based facilities, the most important discoveries about the universe might well be made elsewhere and would most strongly stimulate and inspire young people of other countries.

Optical and Infrared Astronomy

The total area of the mirrors used for astronomical observation is a plausible, if imperfect, figure of merit for assessing the strength in optical and infrared astronomy of the United States relative to other countries. There is every reason to believe that the telescopes that will be built by other countries in the 1990s will be of the same quality as those planned for the United States and that these instruments will be located in comparably good sites. By this standard (Table B.2), Europe and Japan will be making great strides in optical astronomy, with major new instruments coming into operation in the 1990s. However, by the end of the century, U.S. astronomers will still have access to roughly one-third to one-half of the world's telescope area, depending on which private and national telescopes now under consideration are actually built.

Radio Astronomy

The major facilities for millimeter wavelength radio astronomy are discussed in Chapter 3 (see Table 3.2). The major facilities for centimeter wavelength astronomy are described in Tables B.3 and B.4.

TABLE B.2 U.S. Fraction of World Telescope Area for Large (> 3 m) Telescopes

Telescope Name	U.S. Fraction of Time	Mirror Diam. (m)	No. of Mirrors	Date	U.S. Fraction of Telescope Area
Palomar	1	5	1	1950	1.00
Lick	1	3	1	1959	1.00
NOAO-KPNO[a]	1	4	1	1974	1.00
CTIO[a]	1	4	1	1975	1.00
AAT	0	3.9	1	1975	0.81
Calar Alto	0	3.5	1	1976	0.71
ESO	0	3.6	1	1976	0.62
USSR	0	6	1	1976	0.46
IRTF[a]	1	3	1	1979	0.50
UKIRT	0.15	3.8	1	1979	0.47
MMT	1	1.8	6	1980	0.52
CFHT	0.15	3.6	1	1980	0.50
Herschel	0	4.2	1	1987	0.46
ESO NTT	0	3.6	1	1989	0.43
Keck	1	10	1	1991	0.61
ARC	1	3.5	1	1991	0.62
WIYN	1	3.5	1	1992	0.63
MMT-upgrade	1	6.5	1	1994	0.65
Spec. Survey[b]	1	8	1	1995	0.70 (0.65)[c]
VLT-1	0	8	1	1995	0.61 (0.56)[c]
Keck-II[b]	1	10	1	1996	0.67 (0.56)[c]
Columbus[b]	0.75	8	2	1996	0.69 (0.56)[c]
Japan	0.1	7.5	1	1997	0.65 (0.51)[c]
VLT-2	0	8	1	1997	0.60 (0.45)[c]
Magellan[b]	1	8	1	1997	0.61 (0.45)[c]
VLT-3	0	8	1	1998	0.57 (0.40)[c]
NOAO[a,b]	1	8	2	1997	0.62 (0.40)[c]
VLT-4	0	8	1	1999	0.58 (0.37)[c]

[a]Generally available to U.S. community.
[b]Proposed.
[c]Value in parenthesis is U.S. fraction if proposed U.S. telescopes are not built.

TABLE B.3 Centimeter Wavelength Radio Telescopes—Major Filled Aperture Radio Telescopes

Instrument	Diameter (m)	Minimum Wavelength (cm)	Location
MPIR	100	1.3	Bonn, Germany (1970)
NAIC	300	6	Arecibo, Puerto Rico (1963)
NRAO GBT	100[a]	0.7[b]	Green Bank, West Virginia (1995)

[a]Effective projected aperture.
[b]Long-term design goal 3 mm.

TABLE B.4 Centimeter Wavelength Radio Telescopes—Major Radio Interferometers

Instrument	No. of Elements	Element Size (m)	Minimum Wavelength (cm)	Maximum Size (km)	Location
MERLIN	6[a]	25–76	1.3	133	U.K. (1980)
NRAO VLA	27	25	1.3	35	Socorro, N. Mex. (1980)
NRAO VLBA	10	25	0.7	8,000	U.S. (1992)
GMRT	30	45	20	18	Puna, India (1993)
Australia Telescope	6[b]	22	0.7	6	Australia (1990)

[a]Seventh antenna to be in operation in 1991 will extend baseline to 220 km.
[b]Seventh antenna to be added in 1991 will extend baseline to 115 km.

Appendix C

Contributing Scientists

More than 15 percent of all professional astronomers in the United States contributed actively in some way to the astronomy and astrophysics survey for the 1990s. Approximately 300 astronomers served formally on one or more of the 15 panels established by the survey committee to advise the committee on important issues affecting subfields within astronomy. Many hundreds of other astronomers made valuable contributions by supplying written advice to the committee, through their participation at open meetings at the American Astronomical Society or at other professional meetings, or through informal discussions held by committee members throughout the United States.

The reports of most of those panels appear in the separately published *Working Papers* (NRC, 1991) issued simultaneously with this volume. Abbreviated versions of the reports of the Computing and Data Processing Panel and the Status of the Profession Panel appear as Chapter 5 and Appendix B, respectively, in this volume. The report of the Science Opportunities Panel appears only as Chapter 2 and not in the *Working Papers*. The committee also established a working group to examine the potential of lunar astronomy. This appendix lists the scientists who served on panels* or as members of a working group on behalf of the survey.

---

*An asterisk indicates membership in a core group with primary responsibility for writing the panel's report.

BENEFITS TO THE NATION FROM ASTRONOMY AND ASTROPHYSICS

VIRGINIA TRIMBLE, University of California, Irvine; University of Maryland, *Chairman*; JOHN N. BAHCALL, Institute for Advanced Study, *Vice-Chair*; ERIC CHAISSON, Space Telescope Science Institute; ARTHUR CODE, University of Wisconsin, Madison; EDWARD K. CONKLIN, FORTH, Inc.; JOHN COWAN, University of Oklahoma; ALEXANDER DALGARNO, Harvard-Smithsonian Center for Astrophysics; FRANK DRAKE, University of California, Santa Cruz; REBECCA ELSON, Harvard-Smithsonian Center for Astrophysics; GEORGE FIELD, Harvard University; ANDREW FRAKNOI, Astronomical Society of the Pacific; HERBERT FRIEDMAN, Naval Research Laboratory; JOHN S. GALLAGHER, AURA; JESSE GREENSTEIN, California Institute of Technology; HERBERT GURSKY, Naval Research Laboratory; STEPHEN MARAN, NASA Goddard Space Flight Center; PHILIP MORRISON, Massachusetts Institute of Technology; CATHERINE A. PILACHOWSKI, Kitt Peak National Observatory; PHILIP SADLER, Harvard-Smithsonian Center for Astrophysics; CARL SAGAN, Cornell University; PHILIP SCHWARTZ, Naval Research Laboratory; STEVEN SHORE, NASA Goddard Space Flight Center; ALEXANDER G. SMITH, University of Florida; HARLAN SMITH, University of Texas, Austin; MICHAEL S. TURNER, Fermi National Accelerator Laboratory; PAUL A. VANDEN BOUT, National Radio Astronomy Observatory; GART WESTERHOUT, U.S. Naval Observatory; JAMES WESTPHAL, California Institute of Technology; R. STEPHEN WHITE, University of California, Riverside; ALBERT WHITFORD, University of California, Santa Cruz

COMPUTING AND DATA PROCESSING

LARRY SMARR, University of Illinois at Urbana-Champaign, *Chair*; WILLIAM PRESS, Harvard-Smithsonian Center for Astrophysics, *Vice-Chair*; DAVID W. ARNETT, University of Arizona; ALASTAIR G.W. CAMERON, Harvard-Smithsonian Center for Astrophysics; RICHARD M. CRUTCHER, University of Illinois at Urbana-Champaign; DAVID J. HELFAND, Columbia University; PAUL HOROWITZ, Harvard University; SUSAN G. KLEINMANN, University of Massachusetts, Amherst; JEFFREY L. LINSKY, University of Colorado, Boulder; BARRY F. MADORE, California Institute of Technology; RICHARD A. MATZNER, University of Texas, Austin; CLAIRE ELLEN MAX, Lawrence Livermore National Laboratory; DIMITRI MIHALAS, University of Illinois at Urbana-Champaign; RICHARD A. PERLEY, National Radio Astronomy Observatory, Socorro; THOMAS A. PRINCE, California Institute of Technology; CHRISTOPHER T. RUSSELL, University of California, Los Angeles; ETHAN J. SCHREIER, Space

Telescope Science Institute; JOHN R. STAUFFER, NASA Ames Research Center; GERALD JAY SUSSMAN, The MIT Artificial Intelligence Laboratory; JOSEPH H. TAYLOR, JR., Princeton University; JURI TOOMRE, University of Colorado, Boulder; JOHN A. TYSON, AT&T Bell Laboratories; SIMON D.M. WHITE, University of Arizona; JAMES R. WILSON, Lawrence Livermore National Laboratory; KARL-HEINZ WINKLER, Los Alamos National Laboratory; STANFORD E. WOOSLEY, University of California, Santa Cruz

HIGH ENERGY FROM SPACE

BRUCE MARGON, University of Washington, *Chair*; CLAUDE CANIZARES, Massachusetts Institute of Technology, *Vice-Chair*; RICHARD C. CATURA, Lockheed Palo Alto Research Laboratory; GEORGE W. CLARK, Massachusetts Institute of Technology; CARL E. FICHTEL, NASA Goddard Space Flight Center; HERBERT FRIEDMAN, Naval Research Laboratory; RICCARDO GIACCONI, Space Telescope Science Institute; JONATHAN E. GRINDLAY, Harvard-Smithsonian Center for Astrophysics; DAVID J. HELFAND, Columbia University; STEPHEN S. HOLT, NASA Goddard Space Flight Center; HUGH S. HUDSON, University of California, La Jolla; STEVEN M. KAHN, University of California, Berkeley; FREDERICK K. LAMB, University of Illinois at Urbana-Champaign; MARVIN LEVENTHAL, AT&T Bell Laboratories; ROBERT NOVICK, Columbia University; THOMAS A. PRINCE, California Institute of Technology; REUVEN RAMATY, NASA Goddard Space Flight Center; HARVEY D. TANANBAUM, Harvard-Smithsonian Center for Astrophysics; MARTIN C. WEISSKOPF, NASA Marshall Space Flight Center; STANFORD E. WOOSLEY, University of California, Santa Cruz

INFRARED ASTRONOMY

FREDERICK GILLETT, National Optical Astronomy Observatories, *Chair*; JAMES HOUCK, Cornell University, *Vice-Chair*; JOHN BALLY, AT&T Bell Laboratories; ERIC BECKLIN, University of California, Los Angeles; ROBERT HAMILTON BROWN, Jet Propulsion Laboratory; BRUCE DRAINE, Princeton University; JAY FROGEL, Ohio State University; IAN GATLEY, National Optical Astronomy Observatories; ROBERT GEHRZ, University of Minnesota, Minneapolis; ROGER HILDEBRAND, University of Chicago; DAVID HOLLENBACH, NASA Ames Research Center; BOB JOSEPH, University of Hawaii; MICHAEL JURA, University of California, Los Angeles; SUSAN G. KLEINMANN, University of Massachusetts, Amherst; ANDREW LANGE, University of California, Berkeley; DAN LESTER, University of Texas, Austin; FRANK J. LOW, University of

Arizona; GARY MELNICK, Harvard-Smithsonian Center for Astrophysics; GERRY NEUGEBAUER, California Institute of Technology; THOMAS G. PHILLIPS, California Institute of Technology; JUDITH PIPHER, University of Rochester; MARCIA RIEKE, University of Arizona; B. THOMAS SOIFER, California Institute of Technology; PHILIP M. SOLOMON, State University of New York, Stony Brook; PATRICK THADDEUS, Harvard University; DANIEL WEEDMAN, Pennsylvania State University; MICHAEL W. WERNER, Jet Propulsion Laboratory

INTERFEROMETRY

STEPHEN RIDGWAY, National Optical Astronomy Observatories, *Chair*; ROBERT W. WILSON, AT&T Bell Laboratories, *Vice-Chair*; MITCHELL C. BEGELMAN, University of Colorado, Boulder; PETER BENDER, University of Colorado, Boulder; BERNARD F. BURKE, Massachusetts Institute of Technology; TIM CORNWELL, National Radio Astronomy Observatory; RONALD DREVER, California Institute of Technology; H. MELVIN DYCK, University of Wyoming; KENNETH J. JOHNSTON, Naval Research Laboratory; EDWARD KIBBLEWHITE, University of Chicago; SHRINIVAS R. KULKARNI, California Institute of Technology; HAROLD A. McALISTER, Georgia State University; DONALD W. McCARTHY, JR., University of Arizona; PETER NISENSON, Harvard-Smithsonian Center for Astrophysics; CARL B. PILCHER, NASA Headquarters; ROBERT REASENBERG, Harvard-Smithsonian Astrophysical Observatory; FRANCOIS RODDIER, University of Hawaii; ANNEILA I. SARGENT, California Institute of Technology; MICHAEL SHAO, Jet Propulsion Laboratory; ROBERT V. STACHNIK, NASA Headquarters; KIP THORNE, California Institute of Technology; CHARLES H. TOWNES, University of California, Berkeley; RAINER WEISS, Massachusetts Institute of Technology; RAY J. WEYMANN, Mt. Wilson and Las Campanas Observatory

OPTICAL/IR FROM GROUND

STEPHEN STROM, University of Massachusetts, Amherst, *Chair*; WALLACE L.W. SARGENT, California Institute of Technology, *Vice- Chair*; SIDNEY WOLFF, National Optical Astronomy Observatories, *Vice-Chair*; MICHAEL F. A'HEARN, University of Maryland; J. ROGER ANGEL, University of Arizona; STEVEN V.W. BECKWITH, Cornell University; BRUCE W. CARNEY, University of North Carolina, Chapel Hill; PETER S. CONTI, University of Colorado, Boulder; SUZAN EDWARDS, Smith College; GARY GRASDALEN, University of Wyoming; JAMES E. GUNN, Princeton University; JOHN P. HUCHRA, Harvard-Smithsonian Center for Astrophysics; ROBERTA M. HUMPHREYS, University of Minnesota,

Minneapolis; DAVID L. LAMBERT, University of Texas, Austin; BRUCE W. LITES, National Center for Atmospheric Research; FRANK J. LOW, University of Arizona; DAVID G. MONET, U.S. Naval Observatory; JEREMY R. MOULD, California Institute of Technology; S. ERIC PERSSON, Mt. Wilson and Las Campanas Observatories; PETER ALBERT STRITTMATTER, University of Arizona; ALAN T. TOKUNAGA, University of Hawaii; DONALD C. WELLS, National Radio Astronomy Observatory; MICHAEL W. WERNER, Jet Propulsion Laboratory; JOHN McGRAW, University of Arizona, *Consultant*

PARTICLE ASTROPHYSICS

BERNARD SADOULET,* University of California, Berkeley, *Chair*; JAMES CRONIN,* University of Chicago, *Vice-Chair*; ELENA APRILE, Columbia University; BARRY C. BARISH,* California Institute of Technology; EUGENE W. BEIER,* University of Pennsylvania; ROBERT BRANDENBERGER, Brown University; BLAS CABRERA, Stanford University; DAVID CALDWELL, University of California, Santa Barbara; GEORGE CASSIDAY, University of Utah; DAVID B. CLINE, University of California, Los Angeles; RAYMOND DAVIS, JR., Blue Point, New York; ANDREJ DRUKIER, Applied Research Corporation; WILLIAM F. FRY, University of Wisconsin, Madison; MARY K. GAILLARD, University of California, Berkeley; THOMAS K. GAISSER,* University of Delaware; JORDAN GOODMAN, University of Maryland; LAWRENCE J. HALL, University of California, Berkeley; CYRUS M. HOFFMAN, Los Alamos National Laboratory; EDWARD KOLB, Fermi National Accelerator Laboratory; LAWRENCE M. KRAUSS, Yale University; RICHARD C. LAMB, Iowa State University; KENNETH LANDE, University of Pennsylvania; ROBERT EUGENE LANOU, JR., Brown University; JOHN LEARNED, University of Hawaii; ADRIAN C. MELISSINOS, University of Rochester; DIETRICH MULLER, University of Chicago; DARRAGH E. NAGLE, Los Alamos National Laboratory; FRANK NEZRICK, Fermi National Accelerator Laboratory; P. JAMES E. PEEBLES, Princeton University; P. BUFORD PRICE, University of California, Berkeley; JOEL PRIMACK, University of California, Santa Cruz; REUVEN RAMATY, NASA Goddard Space Flight Center; MALVIN A. RUDERMAN, Columbia University; HENRY SOBEL, University of California, Irvine; DAVID SPERGEL,* Princeton University; GREGORY TARLE, University of Michigan, Ann Arbor; MICHAEL S. TURNER,* Fermi National Accelerator Laboratory; JOHN VAN DER VELDE, University of Michigan, Ann Arbor; TREVOR WEEKES, Harvard-Smithsonian Astrophysical Observatory; MARK E. WIEDENBECK,* University of Chicago; LINCOLN WOLFENSTEIN,

Carnegie-Mellon University; STANFORD E. WOOSLEY, University of California, Santa Cruz; GAURANG YODH,* University of California, Irvine

PLANETARY ASTRONOMY

DAVID MORRISON, NASA Ames Research Center, *Chair*; DONALD HUNTEN, University of Arizona, *Vice-Chair*; MICHAEL F. A'HEARN, University of Maryland; MICHAEL J.S. BELTON, National Optical Astronomy Observatories; DAVID BLACK, Lunar and Planetary Institute; ROBERT A. BROWN, Space Telescope Science Institute; ROBERT HAMILTON BROWN, Jet Propulsion Laboratory; ANITA L. COCHRAN, University of Texas, Austin; DALE P. CRUIKSHANK, NASA Ames Research Center; IMKE DE PATER, University of California, Berkeley; JAMES L. ELLIOT, Massachusetts Institute of Technology; LARRY ESPOSITO, University of Colorado, Boulder; WILLIAM B. HUBBARD, University of Arizona; DENNIS L. MATSON, Jet Propulsion Laboratory; ROBERT L. MILLIS, Lowell Observatory; H. WARREN MOOS, Johns Hopkins University; MICHAEL J. MUMMA, NASA Goddard Space Flight Center; STEVEN J. OSTRO, Jet Propulsion Laboratory; CARL B. PILCHER, NASA Headquarters; CHRISTOPHER T. RUSSELL, University of California, Los Angeles; F. PETER SCHLOERB, University of Massachusetts, Amherst; ALAN T. TOKUNAGA, University of Hawaii; JOSEPH VEVERKA, Cornell University; SUSAN WYCKOFF, Arizona State University

POLICY OPPORTUNITIES

RICHARD McCRAY,* University of Colorado, Boulder, *Chair*; JEREMIAH OSTRIKER,* Princeton University Observatory, *Vice-Chair*; LOREN W. ACTON, Lockheed Palo Alto Research Laboratory; NETA A. BAHCALL, Princeton University; ROBERT C. BLESS, University of Wisconsin, Madison; ROBERT A. BROWN,* Space Telescope Science Institute; GEOFFREY BURBIDGE, University of California, San Diego; BERNARD F. BURKE, Massachusetts Institute of Technology; GEORGE W. CLARK, Massachusetts Institute of Technology; FRANCE A. CORDOVA, Pennsylvania State University; HARRIET L. DINERSTEIN,* University of Texas, Austin; ALAN DRESSLER,* Carnegie Observatories; ANDREA K. DUPREE, Harvard-Smithsonian Center for Astrophysics; MOSHE ELITZUR, University of Kentucky; SANDRA FABER,* University of California, Santa Cruz; RICCARDO GIACCONI, Space Telescope Science Institute; DAVID J. HELFAND, Columbia University; NOEL W. HINNERS, Martin Marietta Corporation; STEPHEN S. HOLT,* NASA Goddard Space Flight Center; JEFFREY L. LINSKY,* University of Colorado, Boulder; ROGER F. MALINA, University of California, Berkeley; CLAIRE ELLEN MAX,

Lawrence Livermore National Laboratory; GOETZ K. OERTEL,* Association of Universities for Research in Astronomy; BENJAMIN PEERY, Howard University; VERA C. RUBIN, Carnegie Institution of Washington; IRWIN SHAPIRO, Harvard-Smithsonian Center for Astrophysics; PETER ALBERT STRITTMATTER, University of Arizona; SCOTT D. TREMAINE, Canadian Institute for Theoretical Astrophysics; PAUL A. VANDEN BOUT, National Radio Astronomy Observatory; JACQUELINE H. VAN GORKOM, Columbia University; J. CRAIG WHEELER, University of Texas, Austin; SIMON D.M. WHITE, University of Arizona

RADIO ASTRONOMY

KENNETH I. KELLERMANN, National Radio Astronomy Observatory, *Chair*; DAVID HEESCHEN, National Radio Astronomy Observatory, *Vice-Chair*; DONALD C. BACKER, University of California, Berkeley; MARSHALL H. COHEN,* California Institute of Technology; MICHAEL DAVIS, National Astronomy and Ionosphere Center; IMKE DE PATER, University of California, Berkeley; DAVID DE YOUNG, National Optical Astronomy Observatories; GEORGE A. DULK,* University of Colorado, Boulder; J.R. FISHER, National Radio Astronomy Observatory; W. MILLER GOSS, National Radio Astronomy Observatory; MARTHA P. HAYNES,* Cornell University; CARL E. HEILES, University of California, Berkeley; WILLIAM M. IRVINE, University of Massachusetts, Amherst; KENNETH J. JOHNSTON,* Naval Research Laboratory; JAMES MORAN, Harvard-Smithsonian Center for Astrophysics; STEVEN J. OSTRO, Jet Propulsion Laboratory; PATRICK PALMER,* University of Chicago; THOMAS G. PHILLIPS, California Institute of Technology; ALAN E.E. ROGERS, Haystack Observatory; NICHOLAS Z. SCOVILLE, California Institute of Technology; PHILIP M. SOLOMON, State University of New York, Stony Brook; JILL C. TARTER, NASA Ames Research Center; JOSEPH H. TAYLOR, JR.,* Princeton University; PATRICK THADDEUS,* Harvard-Smithsonian Center for Astrophysics; JUAN M. USON, National Radio Astronomy Observatory; WILLIAM JOHN WELCH,* University of California, Berkeley; ROBERT W. WILSON,* AT&T Bell Laboratories

SCIENCE OPPORTUNITIES

ALAN LIGHTMAN, Massachusetts Institute of Technology, *Chair*; JOHN N. BAHCALL, Institute for Advanced Study, *Vice-Chair*; SALLIE L. BALIUNAS, Harvard-Smithsonian Center for Astrophysics; ROGER D. BLANDFORD, California Institute of Technology; MARGARET E. BURBIDGE, University of California, La Jolla; MARC DAVIS, University of California, Berkeley; DOUGLAS EARDLEY, University of California, Santa

Barbara; JAMES E. GUNN, Princeton University; PAUL HOROWITZ, Harvard University; EUGENE LEVY, University of Arizona; CHRISTOPHER F. McKEE, University of California, Berkeley; PHILIP C. MYERS, Harvard-Smithsonian Center for Astrophysics; JEREMIAH OSTRIKER, Princeton University Observatory; VERA C. RUBIN, Carnegie Institution of Washington; IRWIN SHAPIRO, Harvard-Smithsonian Center for Astrophysics; MICHAEL S. TURNER, Fermi National Accelerator Laboratory; EDWARD WRIGHT, University of California, Los Angeles

SOLAR ASTRONOMY

ROBERT ROSNER, University of Chicago, *Chair*; ROBERT NOYES, Harvard-Smithsonian Center for Astrophysics, *Vice-Chair*; SPIRO K. ANTIOCHOS, Naval Research Laboratory; RICHARD C. CANFIELD, University of Hawaii; EDWARD L. CHUPP, University of New Hampshire; DRAKE DEMING, NASA Goddard Space Flight Center; GEORGE A. DOSCHEK, Naval Research Laboratory; GEORGE A. DULK, University of Colorado, Boulder; PETER V. FOUKAL, Cambridge Research and Instruments, Inc.; RONALD L. GILLILAND, Space Telescope Science Institute; PETER A. GILMAN, National Center for Atmospheric Research; JOHN W. HARVEY, National Optical Astronomy Observatories; ERNEST HILDNER, National Oceanic and Atmospheric Administration; THOMAS E. HOLZER, High Altitude Observatory, National Center for Atmospheric Research; HUGH S. HUDSON, University of California, La Jolla; STEPHEN KEIL, Air Force Geophysics Laboratory; BARRY J. LABONTE, University of Hawaii; JOHN W. LEIBACHER, National Solar Observatory, National Optical Astronomy Observatories; KENNETH G. LIBBRECHT, California Institute of Technology; ROBERT P. LIN, University of California, Berkeley; BRUCE W. LITES, High Altitude Observatory, National Center for Atmospheric Research; RONALD L. MOORE, NASA Marshall Space Flight Center; EUGENE N. PARKER, University of Chicago; REUVEN RAMATY, NASA Goddard Space Flight Center; DAVID M. RUST, Johns Hopkins University; PETER A. STURROCK, Stanford University; ALAN M. TITLE, Lockheed Research Laboratories; JURI TOOMRE, University of Colorado, Boulder; ARTHUR B.C. WALKER, JR., Stanford University; GEORGE L. WITHBROE, Harvard-Smithsonian Center for Astrophysics; ELLEN G. ZWEIBEL, University of Colorado, Boulder

STATUS OF THE PROFESSION

PETER B. BOYCE,* American Astronomical Society, *Chair*; CHARLES A. BEICHMAN,* Institute for Advanced Study, *Vice-Chair*; HELMUT A. ABT,* National Optical Astronomy Observatories; WENDY HAGEN

BAUER,* Wellesley College; GEOFFREY BURBIDGE,* University of California, San Diego; ANITA L. COCHRAN,* University of Texas, Austin; ROBERT DORFMAN, University of Maryland; HUGH HARRIS,* U.S. Naval Observatory; ROBERT HAVLEN,* National Radio Astronomy Observatory; CHRISTINE JONES, Smithsonian Astrophysical Observatory; JEFFREY L. LINSKY, University of Colorado, Boulder; JULIE LUTZ,* Washington State University; LEE G. MUNDY,* University of Maryland; COLIN A. NORMAN, Space Telescope Science Institute; PATRICK S. OSMER, National Optical Astronomy Observatories; JAY M. PASACHOFF, Williams College; BENJAMIN PEERY, Howard University; R. MARCUS PRICE, University of New Mexico; HARRY SHIPMAN,* University of Delaware; JILL C. TARTER,* NASA Ames Research Center; HARLEY THRONSON,* University of Wyoming

THEORY AND LABORATORY ASTROPHYSICS

DAVID N. SCHRAMM, University of Chicago, *Chair*; CHRISTOPHER F. McKEE, University of California, Berkeley, *Vice-Chair*; CHARLES ALCOCK, Lawrence Livermore National Laboratory; LOU ALLAMANDOLA, NASA Ames Research Center; ROGER A. CHEVALIER, University of Virginia; DAVID B. CLINE,* University of California, Los Angeles; ALEXANDER DALGARNO,* Harvard-Smithsonian Center for Astrophysics; BRUCE G. ELMEGREEN, IBM T.J. Watson Research Center; S. MICHAEL FALL,* Space Telescope Science Institute; GARY J. FERLAND, Ohio State University; BRADLEY W. FILIPPONE,* California Institute of Technology; MARGARET J. GELLER, Harvard-Smithsonian Center for Astrophysics; PETER GOLDREICH, California Institute of Technology; ALAN H. GUTH, Massachusetts Institute of Technology; WICK HAXTON, University of Washington; DAVID G. HUMMER, University of Colorado, Boulder; DIMITRI MIHALAS, University of Illinois at Urbana-Champaign; MICHAEL J. MUMMA, NASA Goddard Space Flight Center; PETER PARKER, Yale University; P. JAMES E. PEEBLES,* Princeton University; MALVIN A. RUDERMAN, Columbia University; GREGORY A. SHIELDS, University of Texas, Austin; PETER L. SMITH, Harvard-Smithsonian Center for Astrophysics; SAUL A. TEUKOLSKY, Cornell University; PATRICK THADDEUS, Harvard-Smithsonian Center for Astrophysics; SCOTT D. TREMAINE,* Canadian Institute for Theoretical Astrophysics; JAMES W. TRURAN, JR., University of Illinois at Urbana-Champaign; JOHN WEFEL, Louisiana State University, Baton Rouge; J. CRAIG WHEELER,* University of Texas, Austin; STANFORD E. WOOSLEY, University of California, Santa Cruz; ELLEN G. ZWEIBEL,* University of Colorado, Boulder

UV-OPTICAL FROM SPACE

GARTH ILLINGWORTH, University of California, Santa Cruz, *Chair*; BLAIR SAVAGE, University of Wisconsin, *Vice-Chair*; J. ROGER ANGEL, University of Arizona; ROGER D. BLANDFORD, California Institute of Technology; ALBERT BOGGESS, NASA Goddard Space Flight Center; C. STUART BOWYER, University of California, Berkeley; GEORGE R. CARRUTHERS, Naval Research Laboratory; LENNOX L. COWIE, Institute for Astronomy, University of Hawaii; GEORGE A. DOSCHEK, Naval Research Laboratory; ANDREA K. DUPREE, Harvard-Smithsonian Center for Astrophysics; JOHN S. GALLAGHER, AURA; RICHARD F. GREEN, Kitt Peak National Observatory; EDWARD B. JENKINS, Princeton University; ROBERT P. KIRSHNER, Harvard-Smithsonian Center for Astrophysics; JEFFREY L. LINSKY, University of Colorado, Boulder; H. WARREN MOOS, Johns Hopkins University; JEREMY R. MOULD, California Institute of Technology; COLIN A. NORMAN, Johns Hopkins University; MICHAEL SHAO, Jet Propulsion Laboratory; HERVEY S. STOCKMAN, Space Telescope Science Institute; RODGER I. THOMPSON, University of Arizona; RAY J. WEYMANN, Mt. Wilson and Las Campanas Observatory; BRUCE E. WOODGATE, NASA Goddard Space Flight Center

WORKING GROUP ON ASTRONOMY FROM THE MOON

CHARLES A. BEICHMAN, Institute for Advanced Study, *Co-Chair*; ROBERT W. WILSON, AT&T Bell Laboratories, *Co-Chair*; JOHN N. BAHCALL, Institute for Advanced Study; NOEL W. HINNERS, Martin Marietta Corporation; JEREMIAH OSTRIKER, Princeton University

Appendix D
Members, Commission on Physical Sciences, Mathematics, and Resources

NORMAN HACKERMAN, Robert A. Welch Foundation, *Chairman*
ROBERT C. BEARDSLEY, Woods Hole Oceanographic Institution
B. CLARK BURCHFIEL, Massachusetts Institute of Technology
GEORGE F. CARRIER, Harvard University
RALPH J. CICERONE, University of California, Irvine
HERBERT D. DOAN, The Dow Chemical Company (retired)
PETER S. EAGLESON, Massachusetts Institute of Technology
DEAN E. EASTMAN, IBM T.J. Watson Research Center
MARYE ANNE FOX, University of Texas
GERHART FRIEDLANDER, Brookhaven National Laboratory
LAWRENCE FUNKHOUSER, Chevron Corporation (retired)
PHILLIP A. GRIFFITHS, Duke University
NEAL F. LANE, Rice University
CHRISTOPHER F. McKEE, University of California, Berkeley
RICHARD S. NICHOLSON, American Association for the Advancement of Science
JACK E. OLIVER, Cornell University
JEREMIAH P. OSTRIKER, Princeton University Observatory
PHILIP A. PALMER, E.I. du Pont de Nemours & Company
FRANK L. PARKER, Vanderbilt University
DENIS J. PRAGER, MacArthur Foundation
DAVID M. RAUP, University of Chicago
ROY F. SCHWITTERS, Superconducting Super Collider Laboratory
LARRY L. SMARR, University of Illinois at Urbana-Champaign
KARL K. TUREKIAN, Yale University

Index

A

B

C

D

E

F

G

H

I

O

P

Q

R

S